THE BOOK OF BAMBOO

by

I-Hsiung Ju

Library of Congress No.
82-70338

Third Edition Printed by
The News-Gazette Corporation
20 W. Nelson Street
Lexington, Virginia 24450

TABLE OF CONTENTS

Page

FROM THE DAYS OF THEIR COMMON ORIGIN

An outline of the appreciation of Chinese Calligraphy and Painting 2

THE BOOK OF BAMBOO

Part I Equipment and Supplies

The four treasures of the Chinese scholar....... 22

The use of the tools........ 25

Part II Basic Exercise on Bamboo

Trunk........ 36

Knots or joints........ 43

Horn-of-deer........ 45

The bamboo leaves........ 50

Part III Grouping of Stems, Branches and Leaves

Composition with more than one stem.......... 60

Cluster of leaves........ 73

Diverse "leaf-formations" for a larger and thicker cluster........ 92

The True Essence........ 100

LIST OF PAINTINGS

Page

A Drypoint (Lee Chapel) 5

A Chinese Landscape Painting 6

Pine & Cranes 11

Broken Bamboo 14

Goldfish (detail) 16

Plum 17

Yellow Bamboo 70

Bamboo in the Wind 102

Bamboo in the Snow 103

Old Bamboo 104

After Rain 105

Spring 106

Red Leaves 107

Yellow Bird 108

From the days of their
COMMON ORIGIN

An Outline of the Appreciation
of
Chinese Calligraphy and Painting.

I

From the days of their common origin, Chinese painting and Chinese writing have been allied arts. They use the same equipment and share aims, techniques, and standards.

In the oldest written sources, tracings in the nature of emblems, in connection with divination were used to serve as Words (字). When those words came to mean the language of a poet, they still remained in pictorial forms, but on the other hand, when they came to mean the real "forms" in painting, they retained some of the earlier sense of "semblances" and "emblems" which nicely defined their function in a composition.

For example, the Chinese character Chu (竹) which means bamboo is just an emblem of two groups of bamboo leaves. Ever since the beginning, bamboo has been written and also been painted in the same manner, in other words, a work depicting bamboo is both a painting and a piece of calligraphy.

Bamboo in seal-style

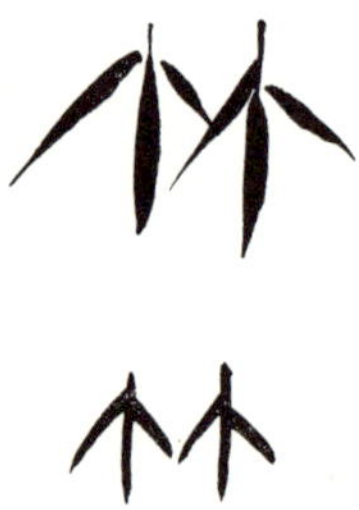

Chu

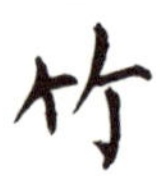

Bamboo in painting

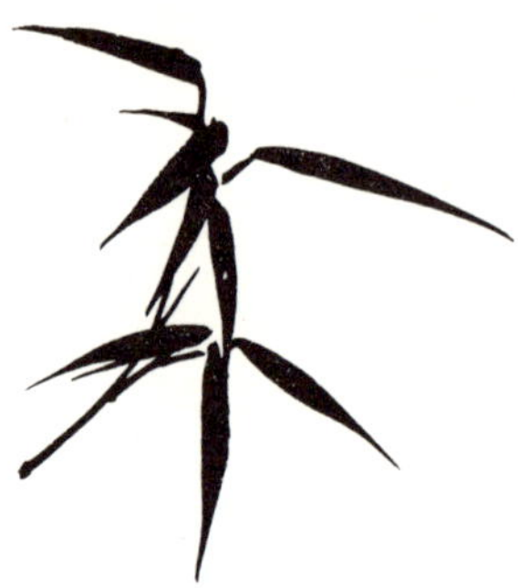

The early Chinese characters had only two "families," one called pictography and another, ideography. They evolved gradually from the Pa Kua (八卦) or Eight Trigrams, invented by **Fu Hsi** (伏羲) which were composed by the grouping together of simple horizontal lines broken and unbroken, symbolizing the diverse objects of the universe.

According to the **I-Ching** (易經) the Book of Changes, the unbroken line is the symbol of Sun, male, brightness – and positive and even anything which has something to do with those mentioned above; the broken line is moon, female, darkness – and passive, and everything that is contrary to the unbroken line.

There are three lines in every group, and eight groups in all, which are

☰	乾 (Chien) Heaven	☵	坎 (K'an) Water
☷	坤 (Kun) Earth	☲	離 (Li) Fire
☴	巽 (Sun) Wind	☶	艮 (Ken) Mountain
☳	震 (Chen) Thunder	☱	兌 (Tui) River

The Eight Trigrams, which were the easiest writing, were also described as first pictures. According to the scratchings on the **Shang** (商) oracle bones and fragments—the prototype of Chinese writing were actually developed from that first picture combinations or diagrams of **I-Ching** (易經).

Ancient books said that the inventor of systematic pictography was **Ch'ang Chi** (倉頡). The following characters are examples:

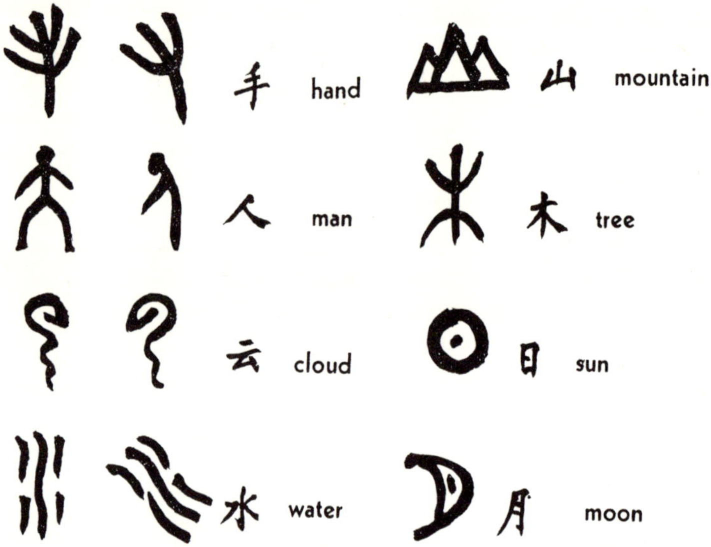

Ancient Chinese characters

In the first "words book" **Shuo Wen Chieh Tse** (說文解字) (literally, explanation and description on words) it was said that the pictographic characters were "drawings of objects by tracing their forms."

We might call every pictographic word an abstract painting. To use the painter's or writer's term, it is Hsie Yi (寫意) meaning — to write ideas. Chinese artists so preferred Hsie Yi (寫意) they had a low opinion of Hsie Shih (寫實) which means to write the real. They enjoyed the symbolic or abstract arts very much, and gave way to "artisans" to work for "realistic" works, which are copies of nature. Chinese painters speak of painting and writing as fundamentally the same, and of styles of brushwork in terms of calligraphy, and they often use the word Hsie (寫) (to write) in place of Hwa (畫) (to paint).

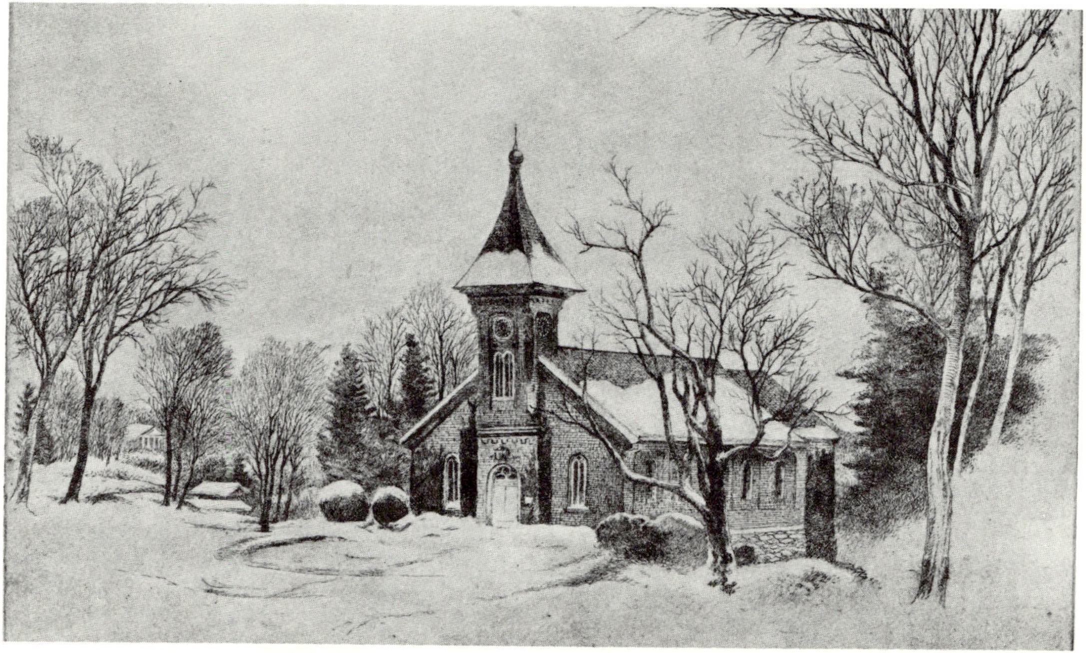

A drypoint done by this author — an art work which "writes" the real —

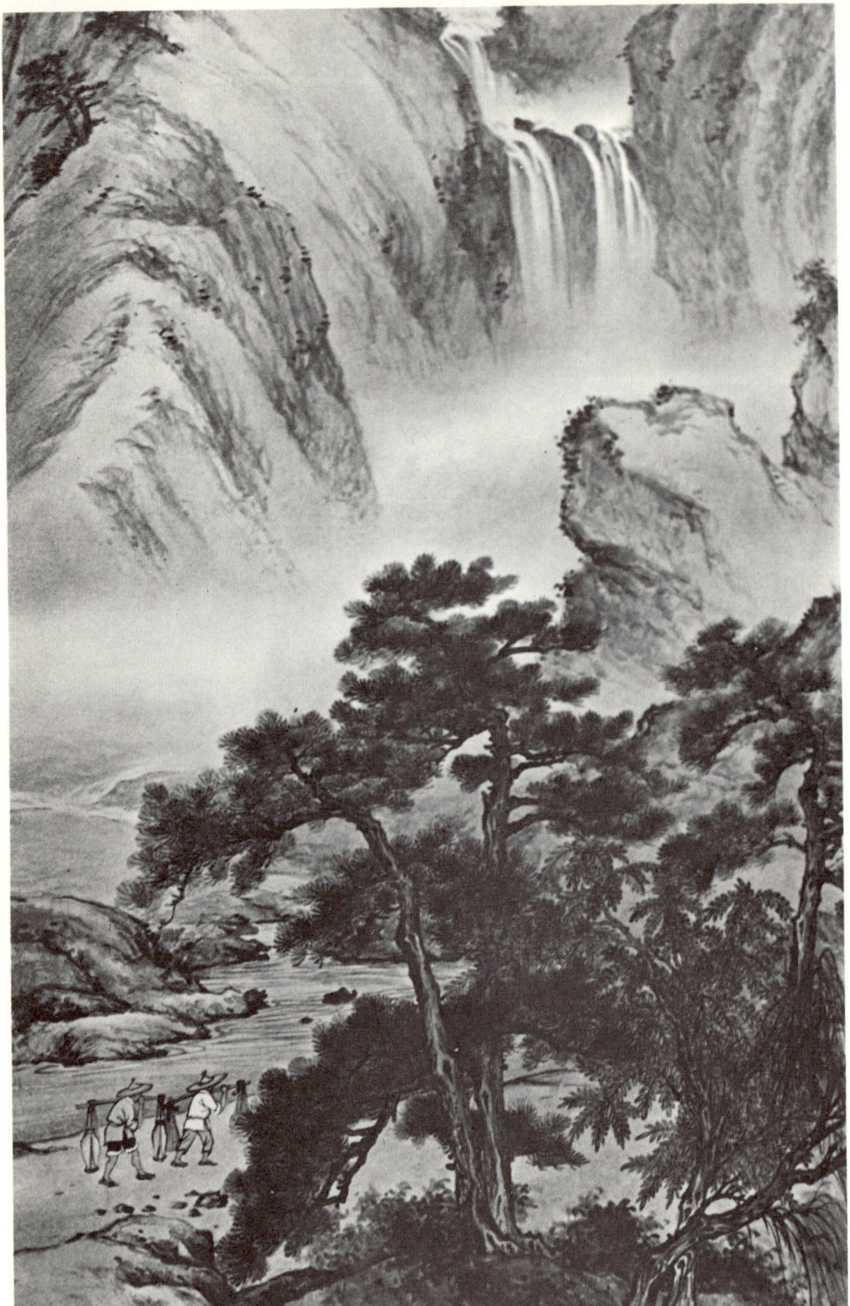

6 A landscape painting done also by this author
— a painting which "writes" ideas —

II

We also found many an ideographic character on the **Shang** (商) oracle bones, **Chou** (周) bronzes, and **Ch'in** (秦) stones. The rubbings from those inscriptions have long been somewhat a treasure to art loving Chinese people.

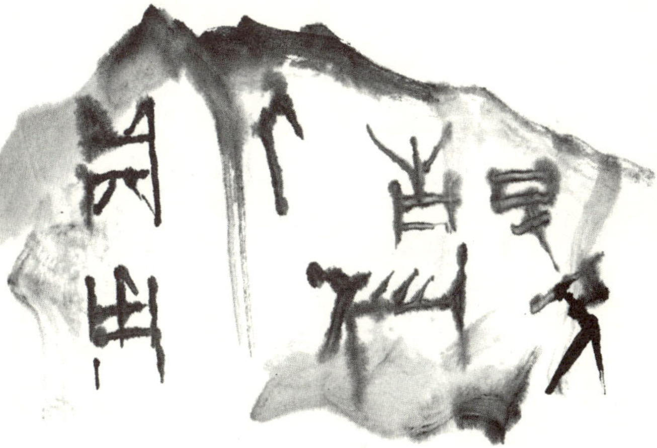

Oracle writings

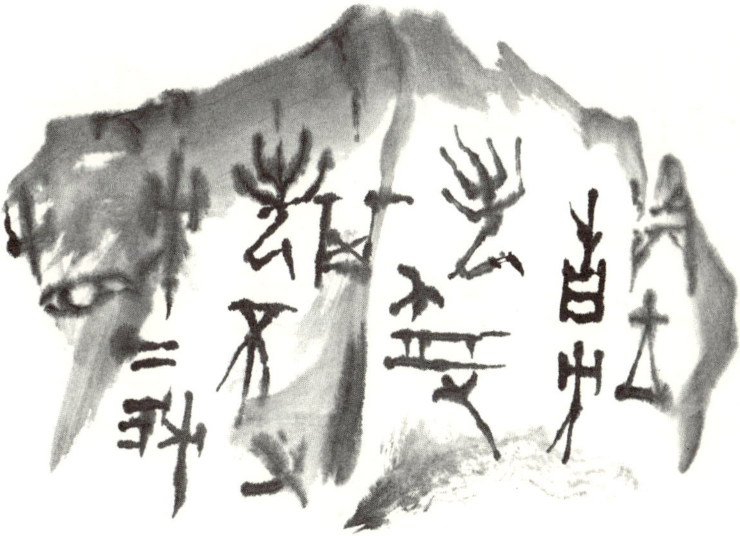

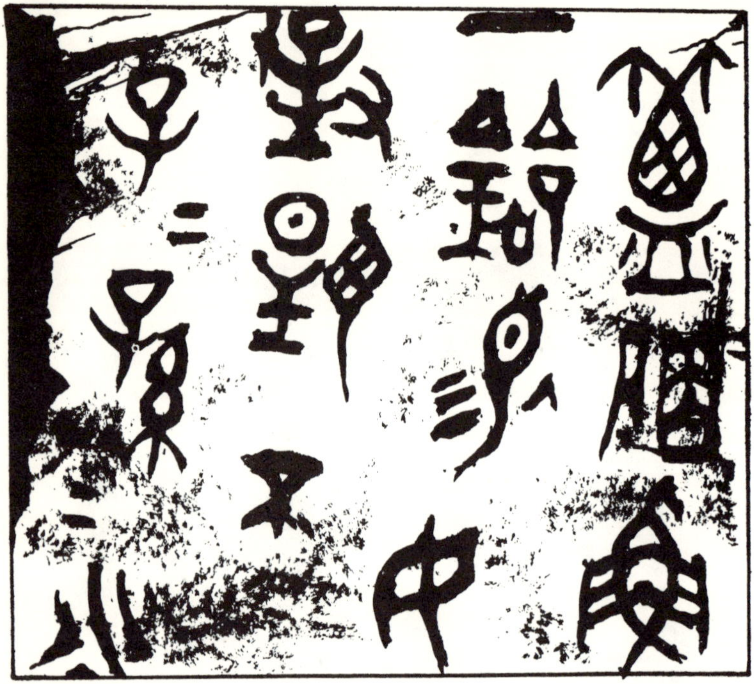

Big-seal style

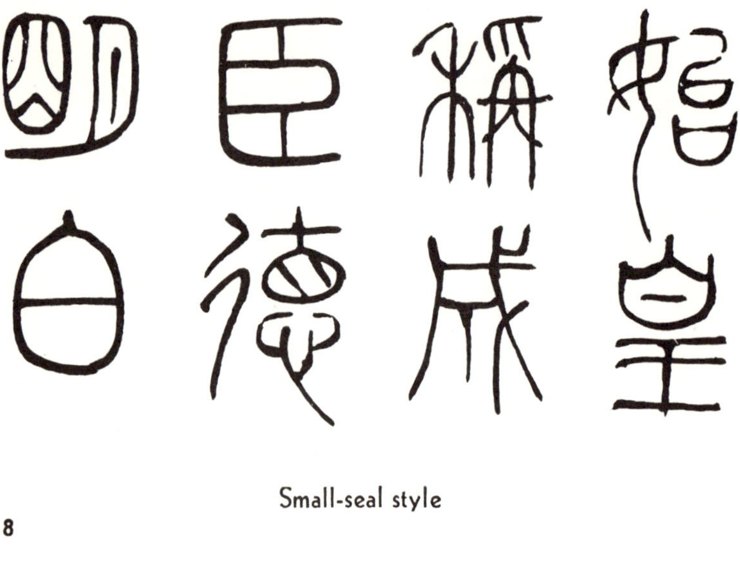

Small-seal style

Ideographic characters have both the idea of Eight Trigrams and of pictography. Chinese Liu Shu (六書) the six origins of Chinese characters, classified ideography into two categories, namely "pointing ideography" and "composed ideography."

For example, using a horizontal line to show the horizon, to put a pointing mark above it means "above," also "on," "up" "top," etc., to put the pointing mark under it makes the word "under," also "below," "down," and "bottom." If we find there is a pointing mark on the upper part of the character "tree," we know it is the "branch," and if the mark is on the lower part, it means "root."

We call this group Chi Shi (指事) literally meaning to point things.

As to composed ideography, we have the symbol of sun above the horizon to mean the word "dawn" or "morning."

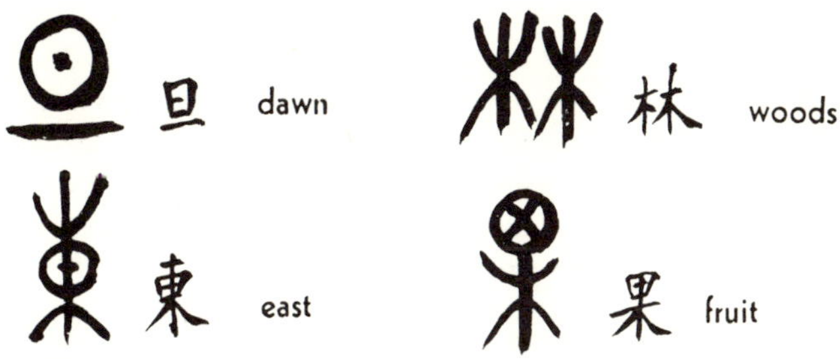

We know the simple picture of "hand," which has only three fingers, can be left-ward or right-ward to serve as the "left" or "right," then an emblem with two hands upward means "offering."

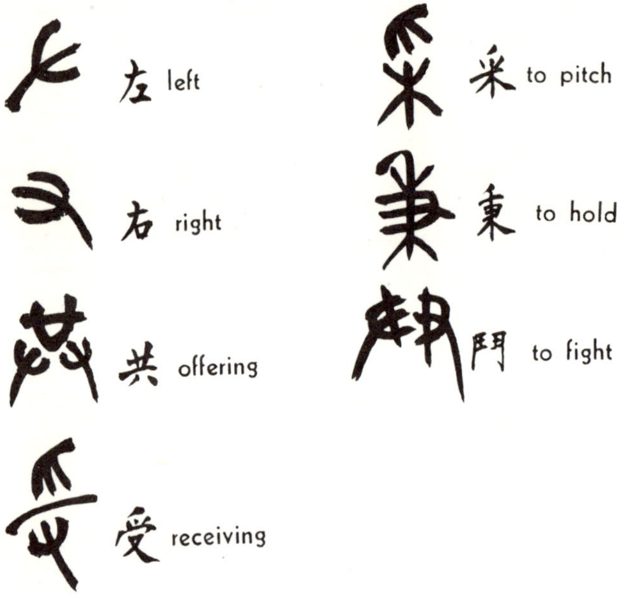

These above characters belong to "composed ideography," which in Chinese is Hwei Yi (會意) meaning "to put the ideas together."

Chinese used three triangles in a line for the word "mountain." We know, words in any language of human beings are expression of the conception. A realistic landscape painting may serve as a good example of mountains, but it is not a fitting symbol to express the conception in the mind.

Chinese artists not only wrote in symbols but also painted in symbols. The attention given to accurate full expression of the artists' idea, is accompanied by an elaborate symbolism. This is a matter of reverence for the traditional interpretations, and a general fondness for an underlying meaning. Since painting is not merely an art for visual appreciation, so we found Chinese masters never had the intention of "grasping" the impression, and so paid no attention to realistic colors.

The idea of layer of symbolic significance is in accordance with the faith in the living quality of bamboo , orchids, trees, and rocks, and all things in nature. Symbolism gives to each living thing a significance beyond the comprehension of its existence as a plant, a tree or a rock.

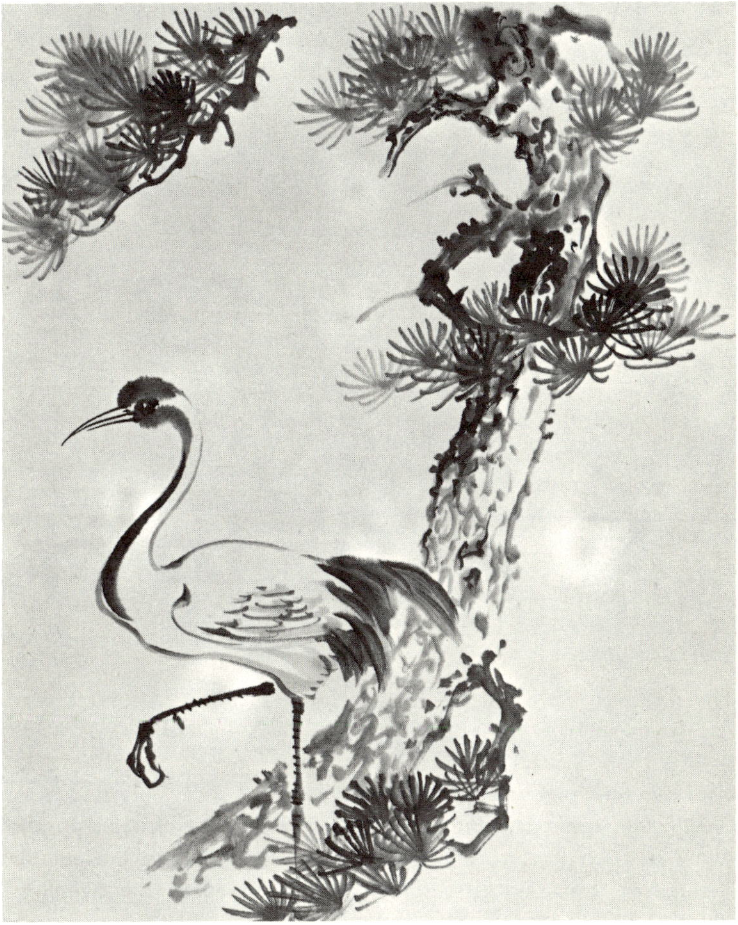

Pine: Because of its evergreen quality, long life, faithfulness
Crane: A bird of mythical powers; messenger of the Immortals; carrier of souls to the Western paradise. Longevity.

III

Now, let us take a look at the following two lines of a poem which is written by a contemporary Chinese artist in Manila. The characters shown here are just a copy from the original and it is in the style of the "oracle bones character."

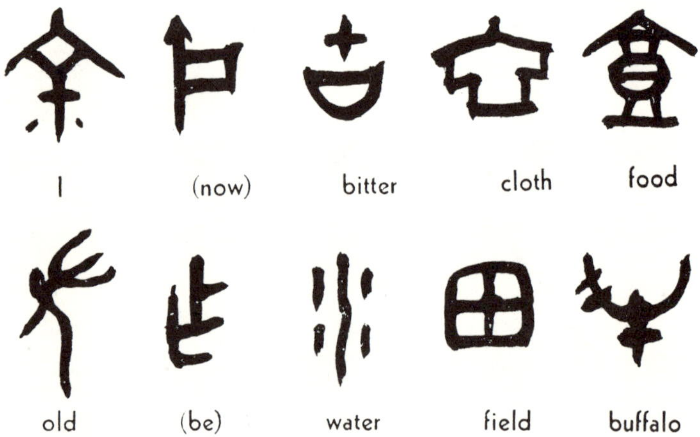

Under the Chinese characters, English words (unrelated words) are given to indicate the meaning of the Chinese characters. These unrelated English words cannot be joined to form the complete sentences, and for this reason they cannot express the meaning of these lines clearly.

The character "field" here is the easiest to be understood, then the "water" reminds you of the K'an (坎) of the Eight Trigrams. What I want to explain more is the construction of the word "bitter" and "old."

To show bitterness, here is an open mouth with a mark of something like a cross above it. To depict oldness, we see a human figure with long "hairs" at the top.

This is what the modern painters – Modernists – are most interested in, for there is nothing that can be recognized as "head" in the character, but "hairs" with utmost emphasis.

Let us forget the meanings. If this is a piece of artwork, or brushwork, can you find anything harmonious among those lines and spaces? Do you think each of them are well constructed and well balanced?

The handling of space, the simplication of brushstrokes to the barest essential, and the symbolism of calligraphy are equivalent to the aims and manners of expression in painting as well as in poetry, which with a few symbolic suggestions imply a wealth of thought and underlying meanings.

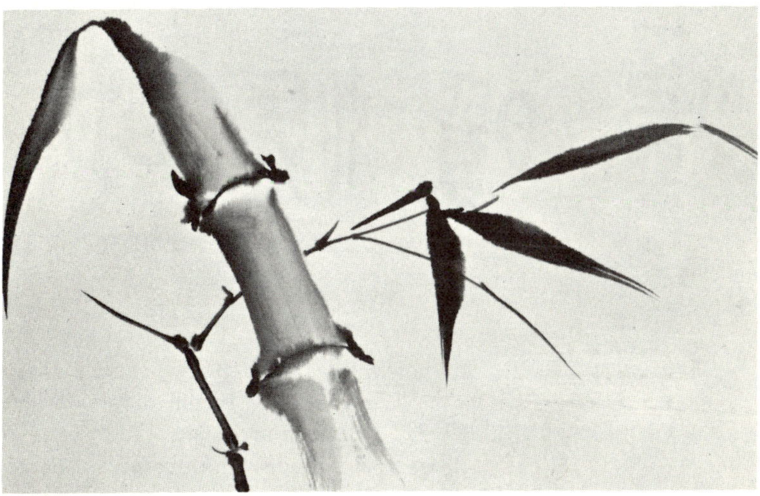

Broken bamboo by I-Hsiung Ju

The following two lines taken from one of the **T'ang** (唐) poet, **Li Shang-Ying's** (李商隱) poem may serve as another sample. The style here is Chuan (篆) which means "Seal."

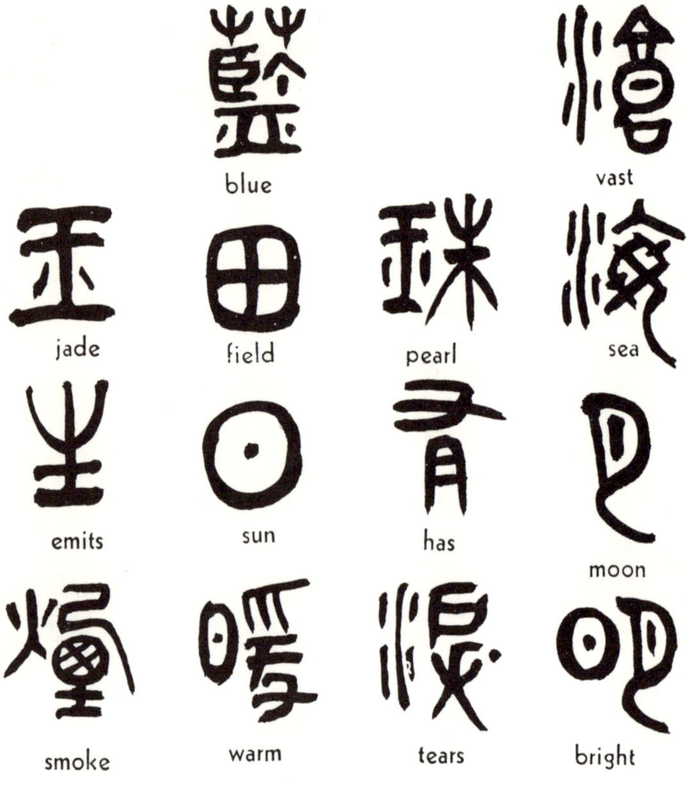

First of all, we see a "couplet." And, please, notice that "vast sea" is contrasted with "blue field," "bright moon" with "warm sun," "pearl" with "jade," "has," with "emits" and "tears" with "smoke." With all these rich allusions which come out only in dreams, the poet was just like a "surealistic" painter trying to give us a colorful painting.

Here again, I introduce you the Ts'ao (草) style of calligraphy, which evolved from the Chuan (篆) Characters.

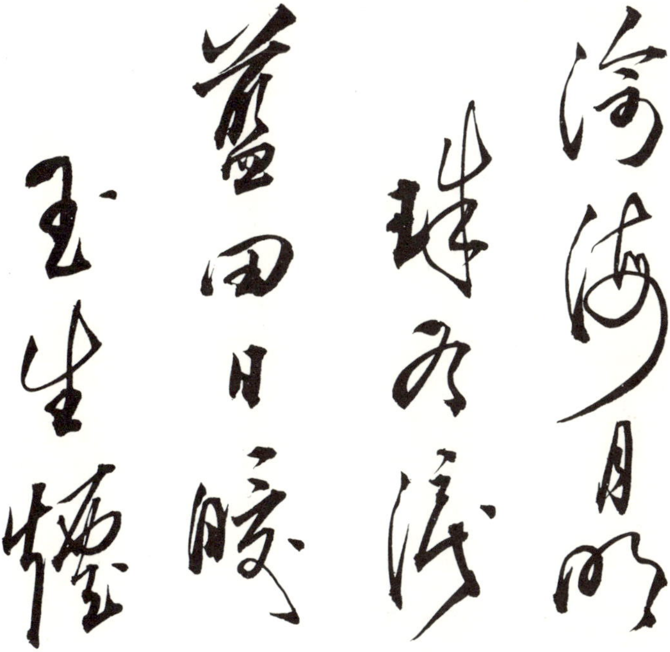

It is the same "couplet" also in vertical order; the first line appear at the right.

The material of Chinese painting – brush, ink, paper, and ink stone— are the same as the simple equipment on the table of a scholar's study. Always held in deep respect, they have long been called "four treasures of the abode of culture," but the brush and ink are most elemental to make a Chinese brush-work attain its characteristic "art-value." When the terms "brush" and "ink" were mentioned in those manuals for the teaching of painting and writing they served not as materials but as techniques. We often have phrases like Yi Tao Pi Bu-Tao (意到筆不到) literally: intention present, brush absent" and Hsi Mo Ru Kin (惜墨如金) literally: "save ink as gold." The Mo (墨) (ink) means "the ink on the paper or silk," or something being written, and the Pi (筆) (brush) here serves as the "method" to handle the brush, and the appearance of brushstrokes.

"Brushstrokes" in Chinese is Pi Ch'u (筆觸) literally meaning: "brush touch." Touch already emits the profound feelings of the artist, for which reason the critics valued the Pi Fa (筆法) i.e., the ways of using the brush. It is a lifelong struggle for an artist to achieve such a level that his brush can be completely controlled to "write" down his "mind" without the least hesitation.

In Chinese painting, the Pi Fa (筆法) or Pi Ch'u (筆觸) of a plumtrunk or a branch of pine is similar to the "big seal," while the bamboo and orchid are close to Ts'ao (草) characters.

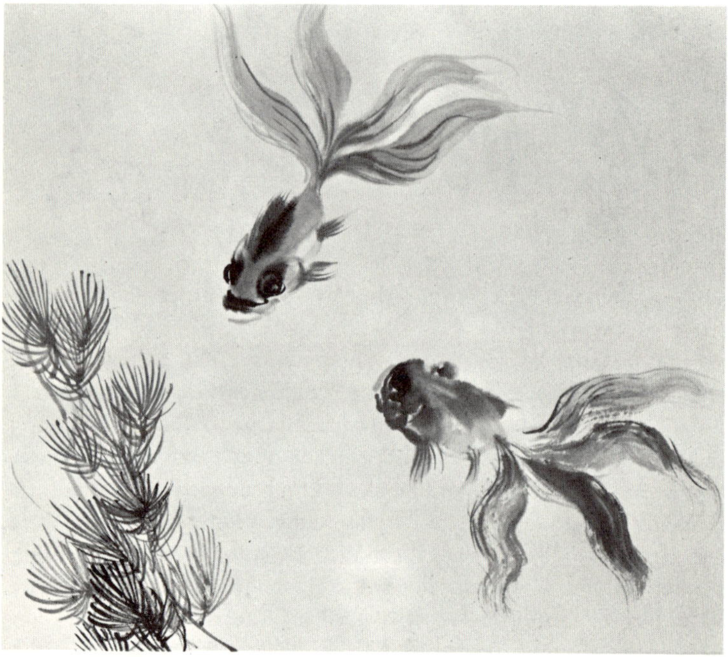

Goldfish (detail) by I-Hsiung Ju

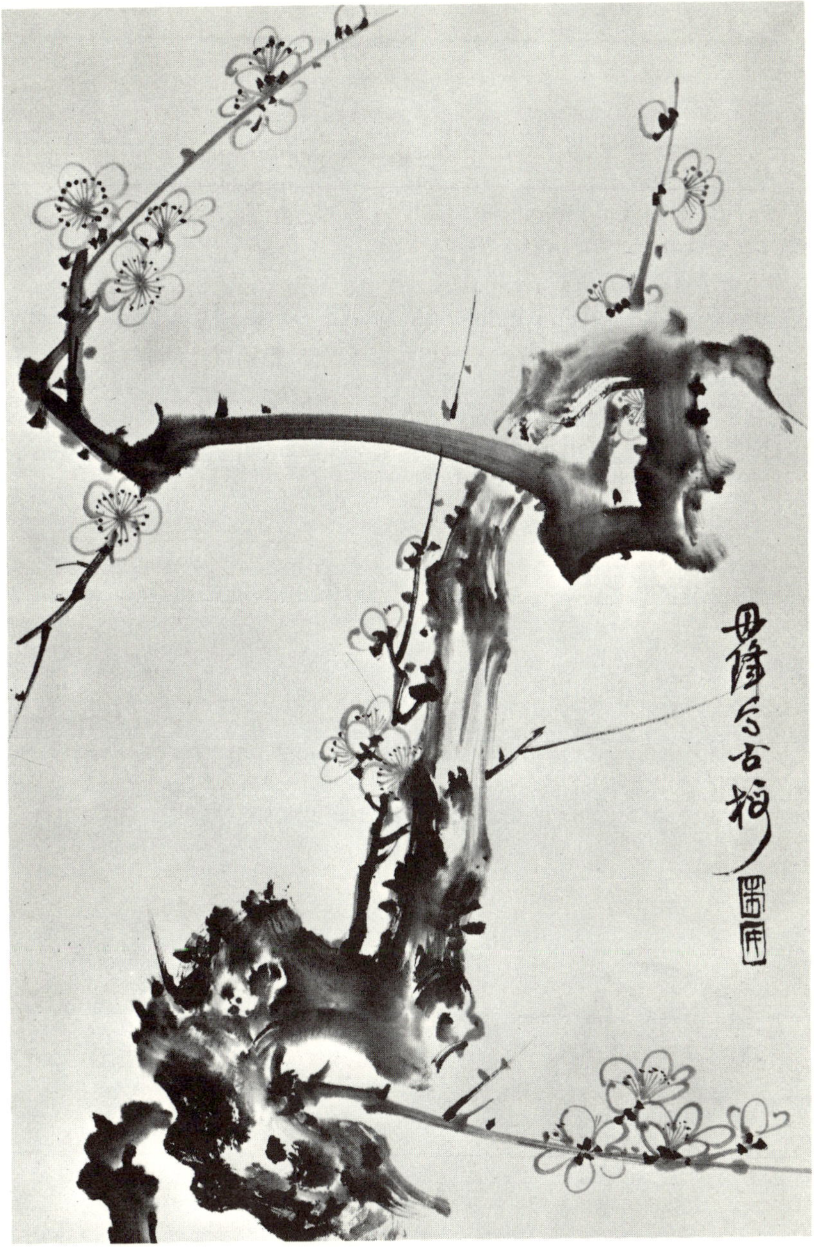

Plum by I-Hsiung Ju

IV

The practice of copying art works plays an important role in the Tao (道) (path) of painting and calligraphy.

Tracing, faithful reproduction, and interpretation of the model are the three "ways" of copying. The purpose for these exercises is to learn brushstrokes as well as methods and certain recurrent designs in composition.

We know that the quality of the brushwork has always been the basis for judging both painting and calligraphy. Attention is directed to the particular degree of pressure of the brush and the manner of applying the ink in either thick or thin strokes, rugged or delicate, light or dark, dry or moist. Special concern must also be taken as to how the strokes are joined. Even the speed and directions of the strokes are carefully studied.

rugged

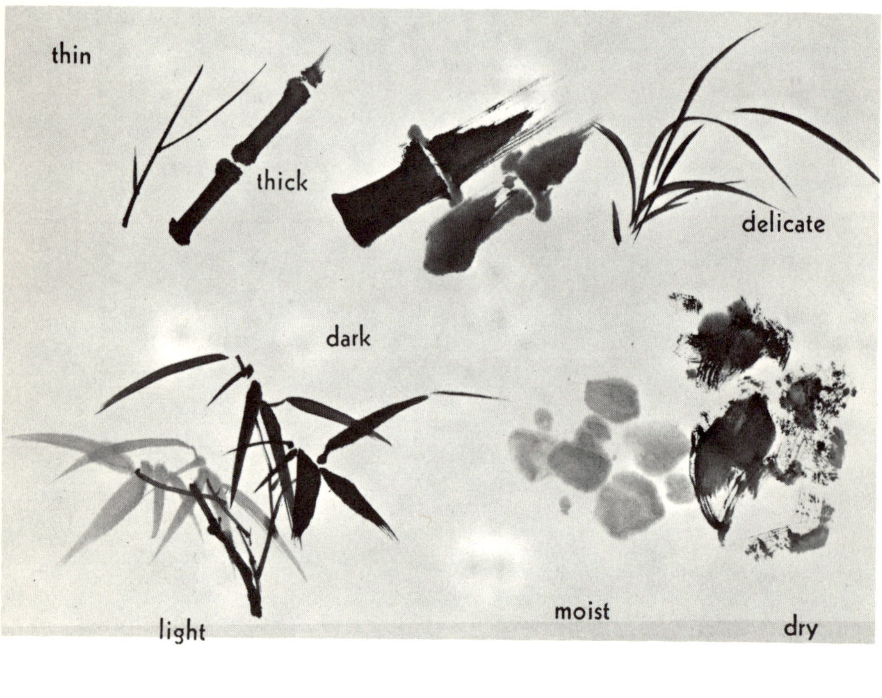

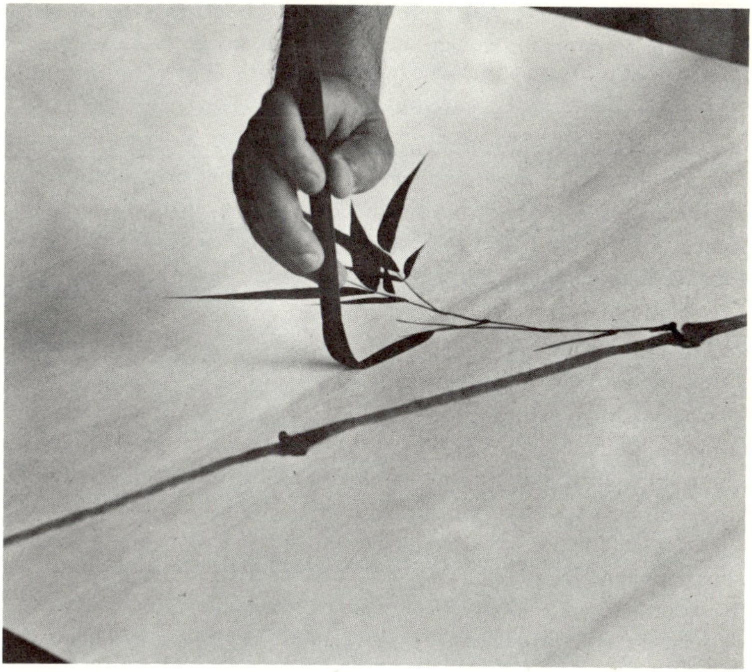

When the artist paints or writes, the brush is held vertically, with no part of the hand touching the paper or silk. Controlled flexibility of the wrist is therefore of great importance. And this ability can be attained only through constant and intensive practice.

The emphasis on brushstrokes in Chinese painting has strengthened the preference for line and the attention given to linear rhythm. Line is the primary means of expressing movement. Line, in relation to space, creates rhythm in the composition. So, the full appreciation of Chinese painting and calligraphy depends a great deal on the spectator's sensitivity to the "tempo" of the brush following it with eyes and imagination as it dots, flicks or moves foreward, sweeping, turning, lifting, plunging, thinning out, swelling, sometimes stopping abruptly, sometimes crouching to leap again.

It has often been remarked that the brush dances and the ink sings.

The nature of the soft brush and loose paper does not permit the slightest correction or changing, or retracing a stroke, without marring the effect of the whole thing. The hand has to be sure; the "heart" and the "mind" must be coordinated.

Only a successful performance can give us wonderful phonograph recordings. Likewise, a well-done brushwork demands a perfect performance. Then, have you ever heard that before the "dance" of the brush and the "singing" of the ink, there are meditations and ceremonies for the artist to keep the "heart" still and the "breath" free?

What is "feeling" to the Western painter is to a Chinese master of the brush a part of the inner vitality.

When strings are touched but faintly by the magical fingers of master violinists, our own feelings are deeply touched and fulfilled. Who but the great pianists can keep the soul of Beethoven ever alive? Have you noticed how delicate and yet powerful are the limbs of those ballet dancers sweeping, turning, lifting, plunging? Theirs is the consumnation of an art, for which they are valued forever. Thus, we understand why Chinese calligraphic works done by masters of consummate art are valued so highly, and why even fragments of ancients rubbings are regarded as rare treasures.

From the **T'ang** (唐) dynasty to the present time, Chinese artists have never stopped to copy and feel proud of their tradition. Maybe, someone would ask: "Isn't it rather monotonous to have only few listed models to be traced or to be followed?"

Yes, it is true. But think also that with the passing of hundreds of years we still have only one Beethoven and one Cezanne. In the history of Chinese art, there are many masters who have left behind them immortal records. Now we have to recognize that all we can hope for is to "transmit" or to "transplant" or, later, to look forward for a break from the tradition.

THE BOOK OF BAMBOO

The first step in Chinese brush painting is learning the technique. This series of books is designed as a step by step procedure to enable the student to learn technique in a more detailed and organized manner. It is a practical handbook for the student of Chinese brush painting.

The first four books in this series will be the "Four Gentlemen" —Bamboo, Orchid, Chrysanthemum, and Plum. The "Four Gentlemen" contain the basic strokes for all Chinese brush painting.

Part I

EQUIPMENT AND SUPPLIES

THE FOUR TREASURES OF THE CHINESE SCHOLAR

The basic tools for Chinese brush painting are the Four Treasures – the ink stone, the ink stick, the brush, and the paper. These treasures are considered as priceless holy things and are treated with reverence and respect.

1. The ink stone is in two parts, the "land" and the "sea." It is usually made of stone, tile or iron.
   a. The "land" or the first part is used to grind the ink.
   b. The "sea" is to hold the ink or the water. The opening is the mouth to pour the ink back to the "land" as it is needed.

   The ink stone is always cleaned with cool water and old ink is removed before the ink stone is put away.

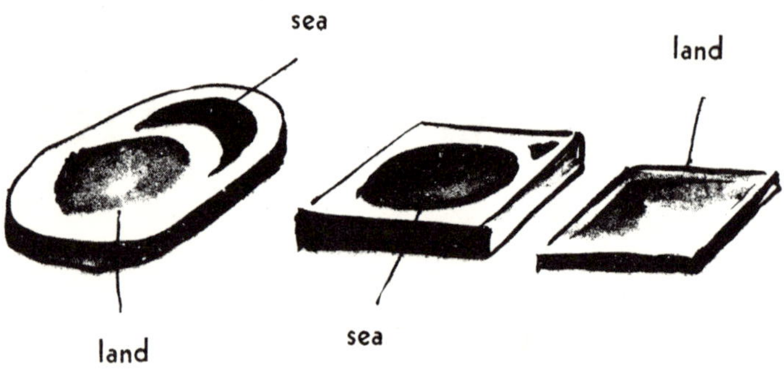

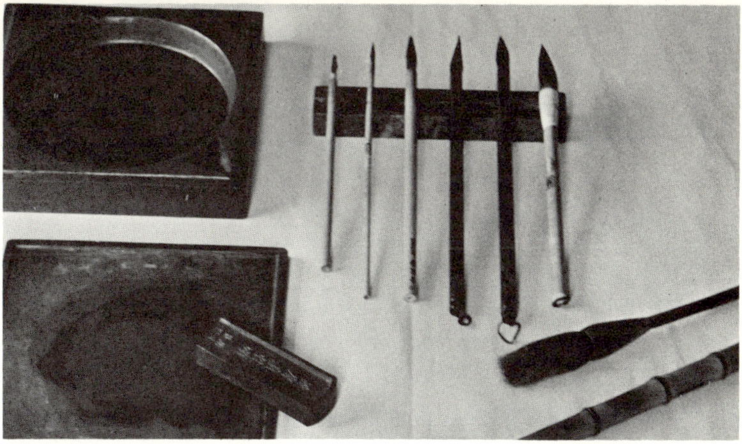

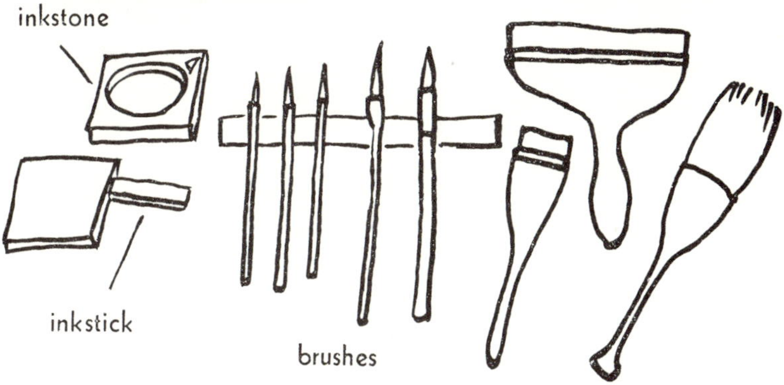

2. The ink stick is a solid cake of ink. There are many different kinds of ink sticks. Prices vary according to sizes, decoration, age, and the maker.

The making of the ink stick and the brush is a carefully guarded secret known to only a few families and passed on only to the sons of the family. At one time all of the families resided in China, but during later days some of the families moved to Japan. Now good ink and good brushes can be found in Japan as well as in China. Generally speaking, ink sticks are made from the soot of burned pine trees or tung trees, mixed with glue and some Chinese ingredients extruded from plants or of rare things such as powdered pearls or amber.

Do not hold the ink stick with wet fingers or let it rest in water because the outside will dry, crack, and break into small pieces.

3. The Chinese brush was invented by General **Meng T'ien** (蒙恬) about 2,000 years ago during the **Ch'in** (秦) Dynasty. From that time the brush was used for writing and painting. The same materials were used for both writing and painting. The picture is considered a book, and the book is considered a picture. The Chinese say, "To write a painting" or "To read a painting."

There are many types of brushes, each with a different use but they are all made by hand with the hairs placed one by one. If the hairs in the brush are black and brown they are from sable, fox, wolf or some animal from the wilds; if the hairs are white they are from the goat.

The student may start with all-purpose brushes which are available in small, medium, and large sizes. Later he will discover that hard hairs are good for lines and for painting trees, soft ones are good for color and for painting the sky. A long, springy brush is good for painting water, a wider brush for painting leaves, and thin, long hairs for painting veins. Also there are special brushes for wash and for pasting.

Brushes are always kept clean and straight.

4. Rice paper is the paper commonly used for Chinese brush painting. Rice paper comes in various textures according to the process used by the manufacturer. If the paper is sized it is less porous and the student must move the brush more slowly than on unsized paper.

Inexpensive newsprint is excellent for the beginner.

A painting on rice paper should be pasted to another piece of rice paper or heavier paper to preserve it.

THE USE OF THE TOOLS

HOW TO GRIND THE INK

The time devoted to the ink is very important, it is a time for meditation. It is a time to calm oneself, to forget every-day cares, and to find the true self.

The first step is to bring water to the "land." Each Chinese brush has a bamboo cover. The cover should never be replaced over the bristles of the brush once the brush is used because it would injure the delicate bristles. However, the cover is an excellent vessel to drop water onto the "land" if it is used in the following way.

With a sharp, pointed instrument puncture and remove the joint of the bamboo which is the top of the cover. Now you have an open bamboo tube. Place the tube halfway into the water. Cover the top of the tube with your finger to form suction of air to hold the water in the tube. Move the tube to the "land," remove your finger from the top opening, and release a drop or two of water.

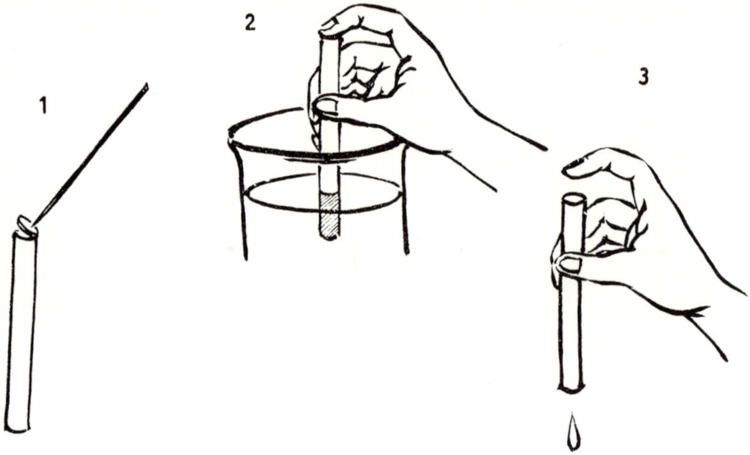

(1) Remove the joint of the bamboo which is the top of the brush cover
(2) Place the tube halfway into the water
(3) Release drops of water to the "land"

The second step is the actual grinding of the ink or the making of liquid ink from the solid ink stick. With dry fingers hold the ink stick upright to the "land" stone. In circular motion grind or rotate the ink stick ten times clockwise, and ten times counter-clockwise. Keep repeating, keep grinding, keep counting. You are meditating and relaxing, and the water is becoming black ink.

The Test for Good Black Ink

When you try to lift the ink stick and feel that the water has become sticky and creamy, you have good black ink. If you need additional ink, pour the first ink into the "sea." Then with several drops of water on the "land" repeat the the same process and grind more ink. After grinding the ink, place the ink stick aside by itself to dry.

If the ink appears to dry on the "land", add a drop of water or use a wet brush to dissolve the ink. Only the water has evaporated; the ink remains.

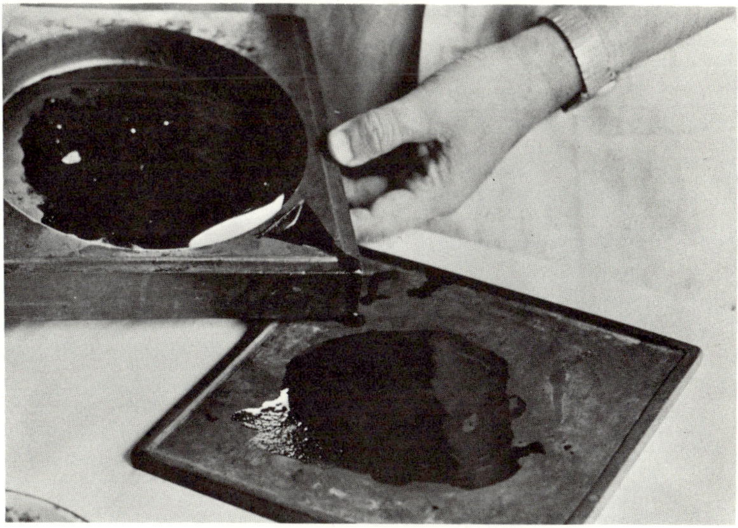

Transferring water from the "sea" to the "land"

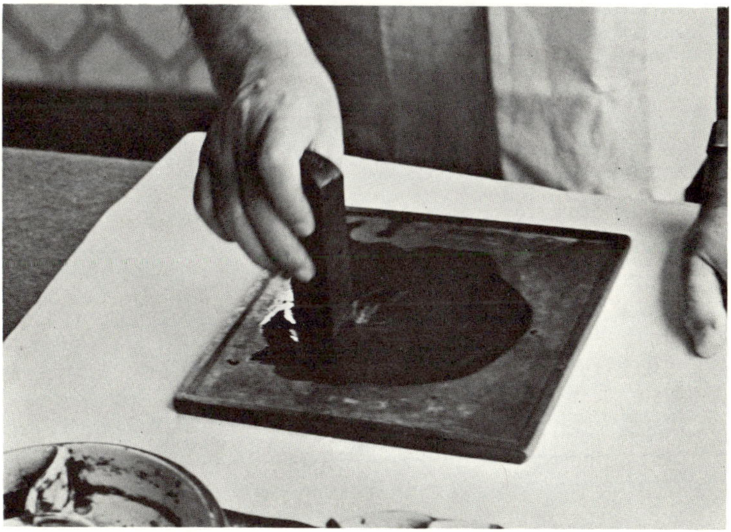

The grinding

Good ink never contains small particles of powder or sand. We must prepare the ink every time we paint. Old ink is subject to chemical reactions and prevents us from getting beautiful gradations of gray.

For gradations of gray we add different amount of water.

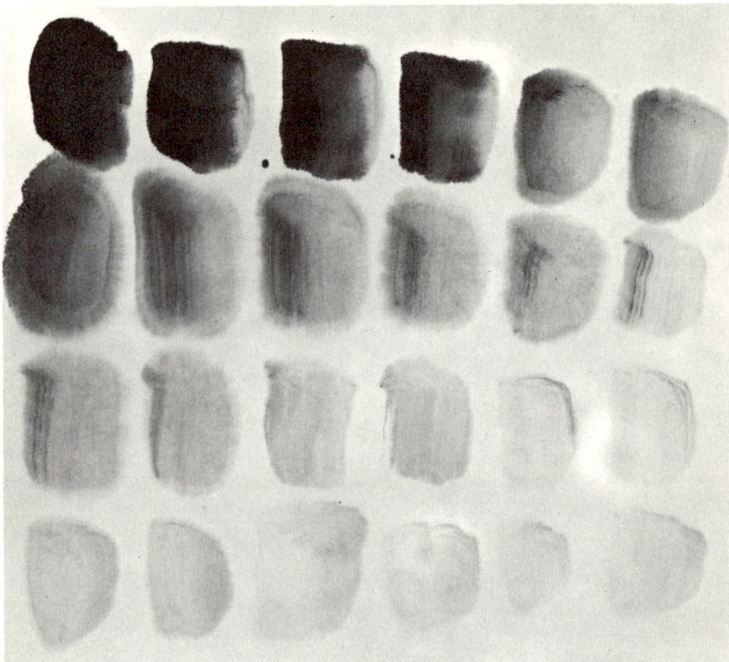

Gradations of gray

To Remove Black Ink from Clothing

Chinese ink is very difficult to remove, but can be removed successfully if the garment is washed immediately.

Make a paste of starchy rice which still has liquid in it. Spread the paste on the fabric and rub into the ink spots. The ink will be absorbed into the starch of the rice. Rinse completely with clean water.

CARE OF THE BRUSH

When the brush is new, remove the bamboo cover and place the brush in cold water to soak until bristles are completely soft. Never replace the cover on the brush after bristles have been soaked because you will damage the brush.

The hairs in the brush should always remain straight and pointed if you hold the brush correctly. If the hairs do become twisted, you may straighten them by gently combing them with an old tooth brush or a pin or needle.

Never press too hard with the brush; never use the brush to clean the stone; and never let ink dry on the hairs. Store brushes flat and rolled in a bamboo mat such as the bamboo mats used for table mats.

HOLDING THE BRUSH (Three Positions)

Remember that the tip of the brush is always straight and pointed if you hold the brush properly. The index finger and the middle finger are on the outside of the brush; the thumb and the ring finger are on the side toward your body.

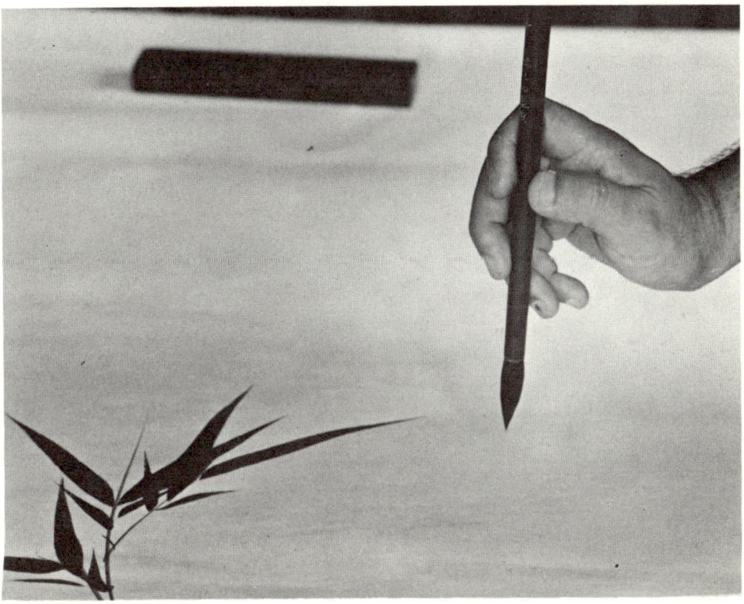

Holding the brush

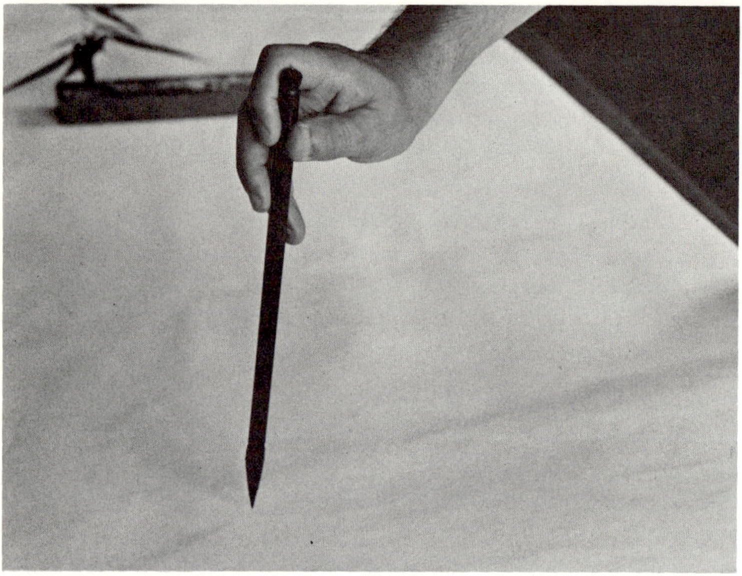

For a line longer than one foot

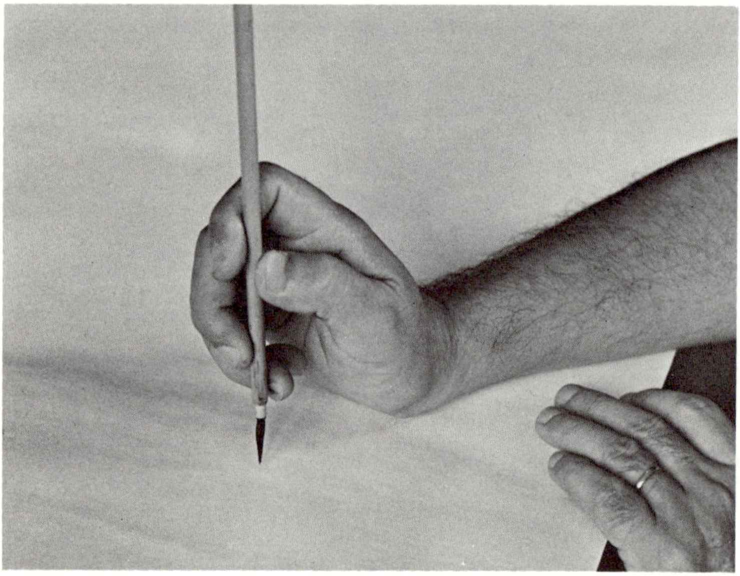

For a movement within a square inch

Another way to explain the holding of the brush would be to say that the brush is held upright between the thumb and two middle fingers.

To get different length of lines there are three positions

a. For a line longer than one foot, hold the brush upright to the paper or table. The arm from the elbow to the fingers forms a right angle to the handle of the brush. No part of the arm touches the table or the paper, and it is completely free to make line longer than one foot without moving the body.

b. To make a line within one foot in length, hold the brush upright to the paper but rest your elbow or forearm on the table.

c. For movement within a square inch, hold the brush upright but rest the heel of the palm on the table. The movement will be with the two fingers and the thumb.

   This position with the finger movement will be used when you write a character, paint small flowers, make the eyes of a bird, etc.
   Sometimes the small finger is used as a guide or as a stopper to keep the brush upright.

TAKING INK FROM THE INK STONE

After the hairs of the brush have softened, remove excess water by pressing slightly on the edge of the water container

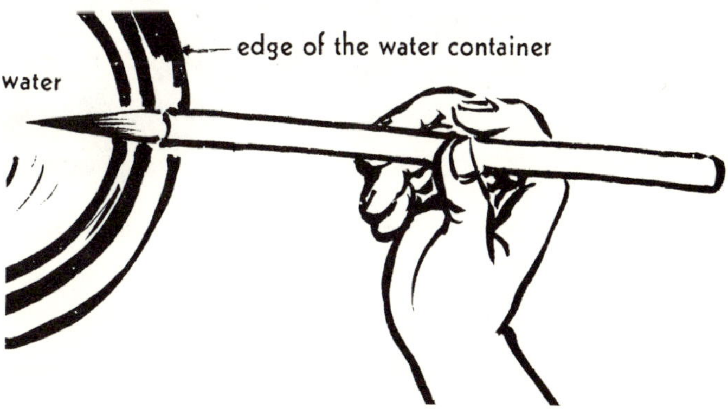

with the brush and pulling it away. With the handle of the brush parallel to the edge of the container you also roll the brush slightly forward and slightly backward. So there are three different kinds of action pressing, pulling, and turning, to remove excess water. The tip of the brush remains straight.

Take up ink from the inkstone with the brush, holding brush horizontal to the table. Pull away from the stone as you press slightly and roll the brush forward and backward to keep the tip pointed and the hairs untangled.

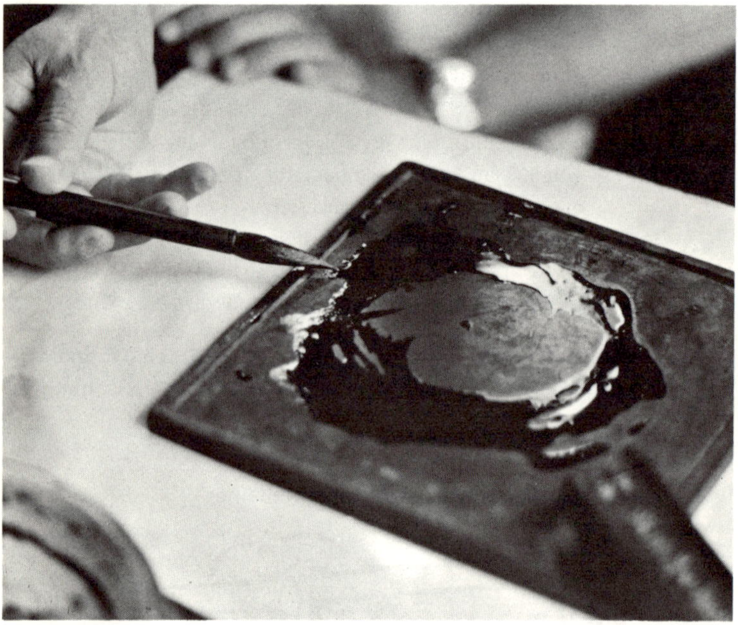

The way to keep the tip pointed and the hairs untangled

To check the amount of ink taken, hold the brush upright. If there is a drop of water or ink, the brush is too wet. If brush is too dry, you must add ink or water. If the tip is perfectly pointed, the brush is just right.

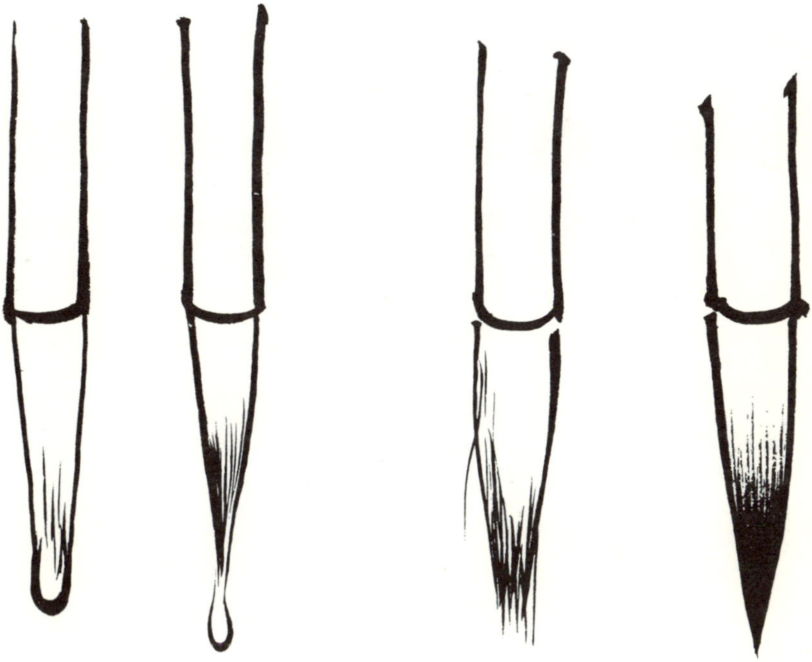

too wet wet too dry just right

Part II

BASIC EXERCISE ON BAMBOO

TRUNK

There are so-called "bamboo painters" who all their lives paint only bamboo. The bamboo is strong, upright, and dependable. He may bend with the wind, the storm and the rain, but he never breaks. He is a true gentleman of courage and endurance.

The first lesson deals with the bamboo because the trunk, the leaves, and all the lines are comparatively straight. There are no large curves or circles.

Practise each step many times before attempting to paint a complete bamboo. The first few days just practise the trunk, later the horn-of-deer, and lastly leaves.

A. **Brush**—the all purpose brush may be used for all bamboo strokes.

B. **Ink and Color**

1. The beginner must paint in black and gray to learn the strokes. Dip brush first in the water. Then take black on the tip and your brush has black, gray and water from the tip up.

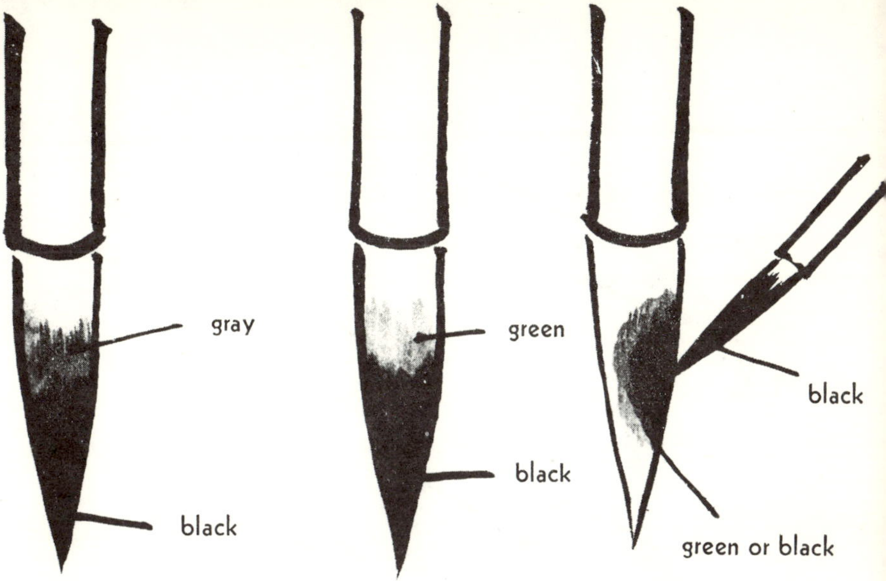

2. Green with black on the tip of brush makes a beautiful stem.
3. Yellowish color indicates older stems and leaves,
4. Stems may be drawn with a stiff brush using dark ink, then painted over with color as a wash.

C. **Procedure and Stroke**

1. The first section. Start from the bottom of the paper and go up. Older sections of bamboo at the bottom of the stem are harder, heavier, and wider, but not very long.

   A "pause" m :ans to press the brush. Pause longer for shorter strokes.

   Each section will resemble the Roman I. This is the first section. Leaving a very small space, move upward for the second section. Holding the brush upright let the tip of the brush touch the paper, push to the left side a bit and then pull back to the right side. This indicates the size of the stem section. Then push the brush back to the middle that bottom line..... pull upward ..... push to the left side and back to the right side and lift the arm from the paper. You have a section of bamboo.

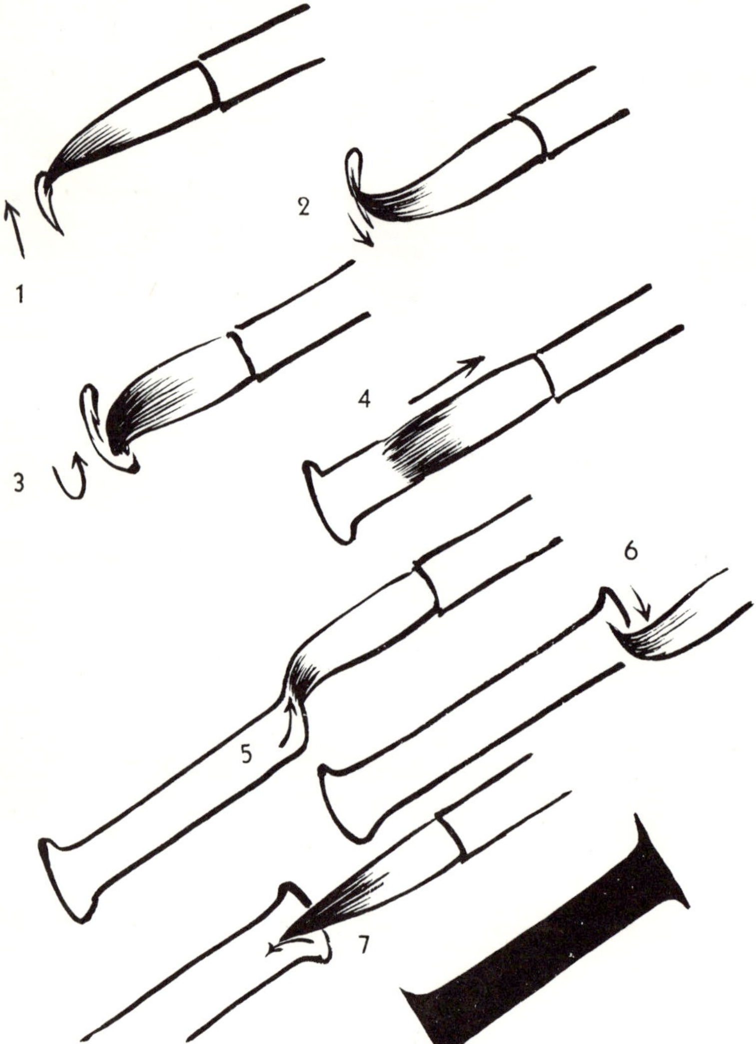

a Roman I
The first section

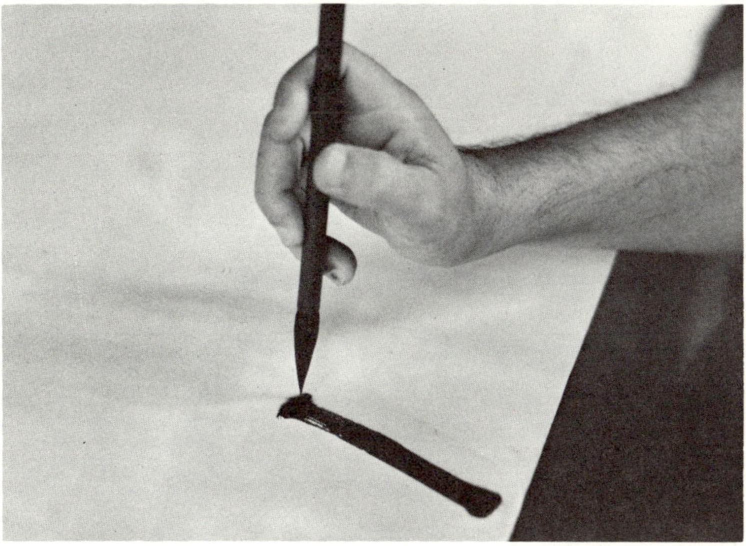

2. The second section. This is almost the same as the first section, but a little bit longer. The pressure is a little bit lighter to make the section slightly narrower than the first.

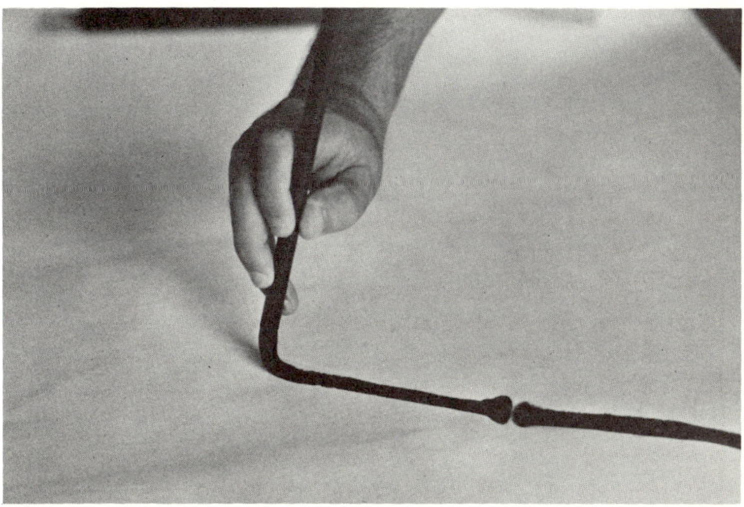

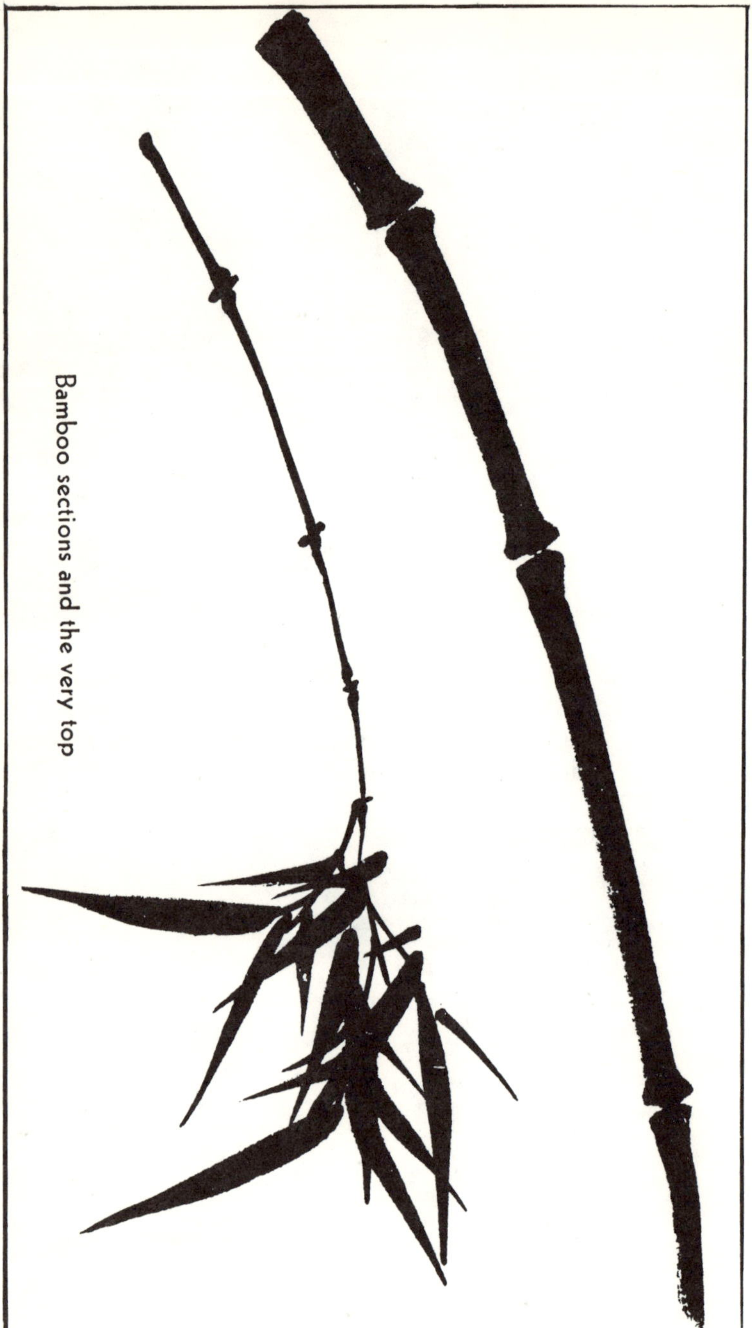

Bamboo sections and the very top

3. The third section, etc. Pressure becomes lighter and the section become longer as the stem grows.

4. Toward the top the bamboo sections become shorter again. You move firmly and steadily. For the first section you press very hard, for the second section you press lighter, for the third section, you press even lighter. The lighter the pressure, the narrower the stem becomes and you must remember that the bamboo stems become thinner as they grow upward. Also each bamboo section becomes a little longer than the one below it, except near the very top of the bamboo the section gradually becomes shorter.

Note: In practice, do not make adjacent stems with joints of the same length.

To make a narrow bamboo, use the tip of the brush only and very light pressure. To make the bamboo stem smooth on both sides keep the handle of the brush straight up . . . . .do not let the handle of the brush drop.

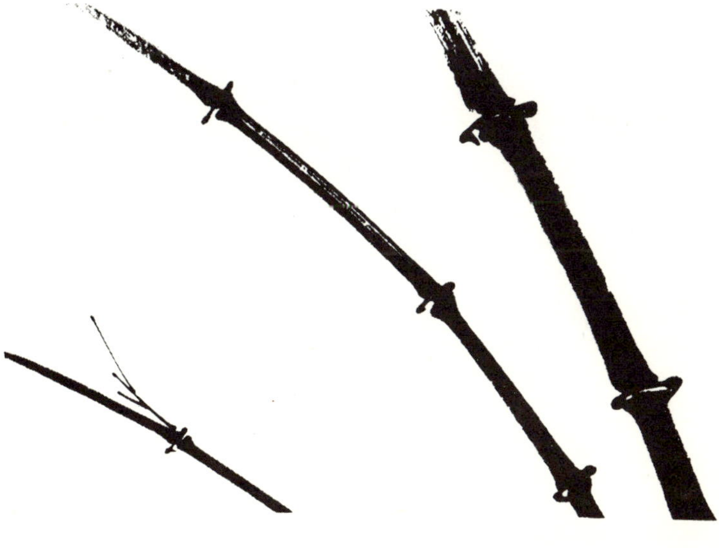

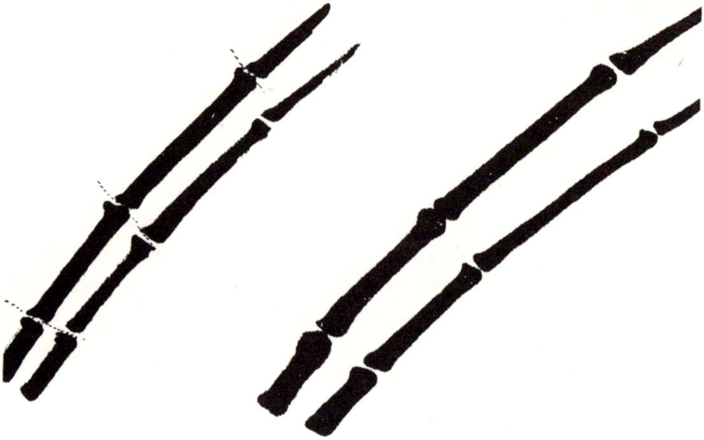

Do not make adjacent stems with joints of the same length

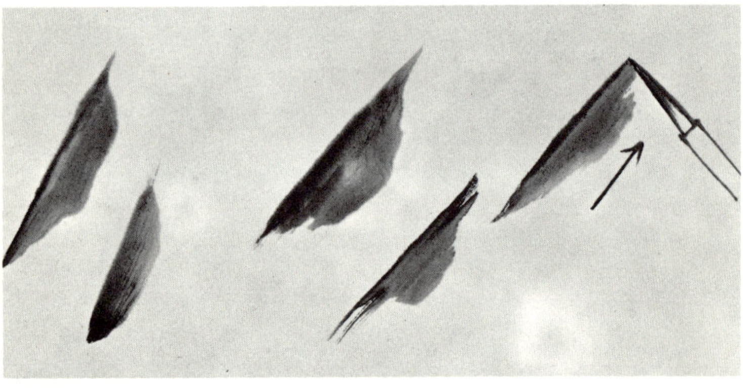

Do not let the handle of the brush drop

KNOTS OR JOINTS

A. The knots or joints are painted with black ink between each bamboo section, from bottom to top of the stem. In this way the joints become lighter in color toward the top of the stem.

The stroke for a joint is a pause . . . a curve to the right . . . and a pause. Hold your brush upright to the paper, keep the same pressure, and never let your brush drop. Let the curve fit the size of the trunk. Avoid making it too short or it will look like a sugar cane.

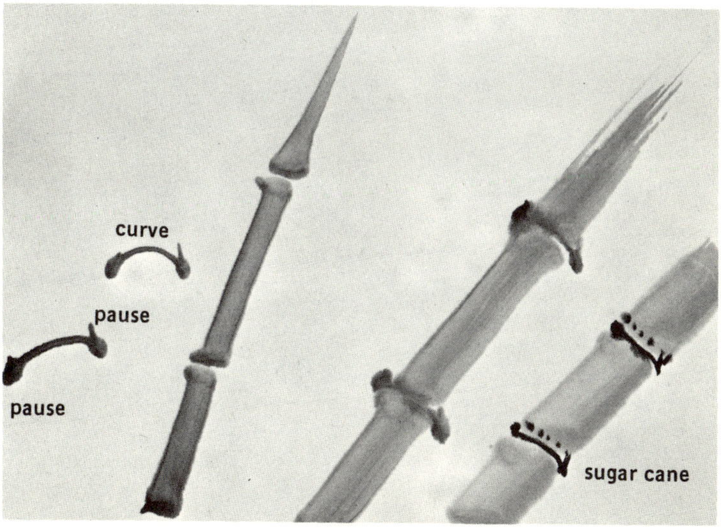

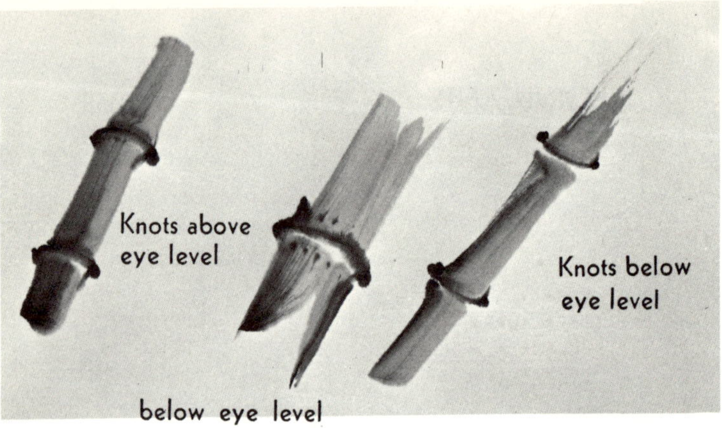

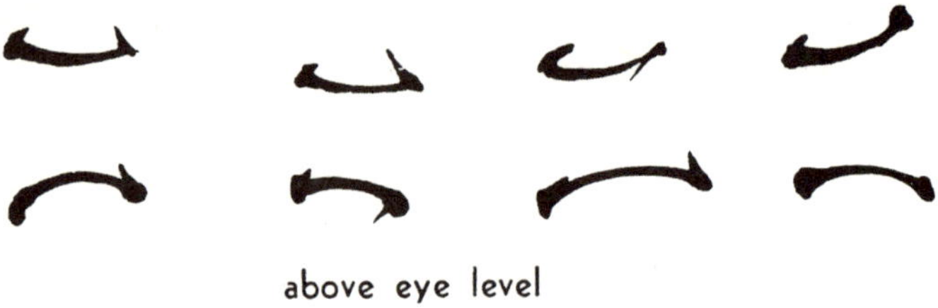

The arm is free and the brush is moving fast. The curve is important and must still appear clearly.

B. The knots or joints must be painted without re-inking the brush. Because the brush is moving downward, the last "pause" stroke will be downward.

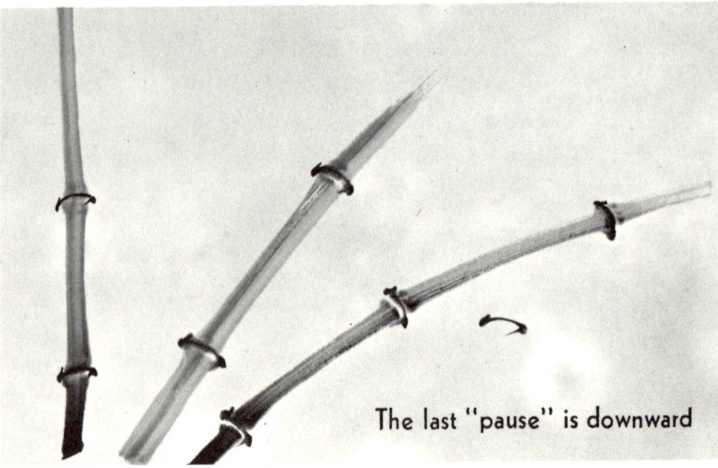

HORN-OF-DEER

The third part of the bamboo is called the horn-of-deer because it looks like the deer's horn or small branches.

The important thing about this stroke is to feel the movement. The Chinese artists prefer to have a good stroke even without ink. That means that the performance is more important than the picture.

Each horn-of-deer has three strokes which can be arranged in different ways.

A. **The Stroke** — begins with a pause and gradually fades away. Hold the brush upright, let the tip touch the paper, press gently, (your hand is moving like a swing), gradually, gradually, slowly lift the brush up, up, up until it is away from the paper. Each line starts with a pause and ends with a point. Remember, for shorter strokes, a longer pause.

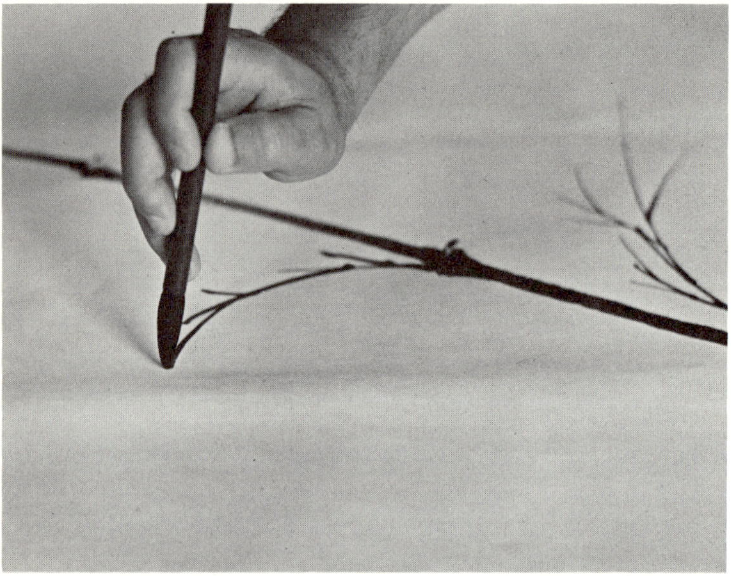

Pause — continue stroke — lifting toward the end
Note: Some students like to use the little finger as a guide.

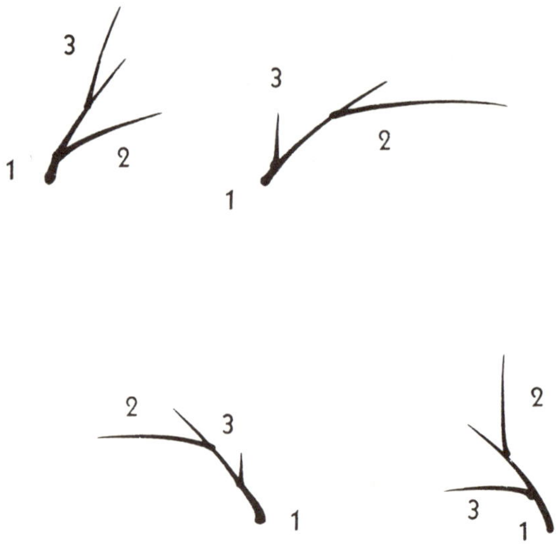

B. **Three ways of placing the stroke** to form horn-of deer are:

1. In three short segments; one in the middle, one on the left side, and one on the right side. The strokes on the left and right sides alternate or stagger; it does not matter which one is higher or lower. They do not start from the same point. If the first one starts on the left side of the middle segment, then the next one will be on the right side.

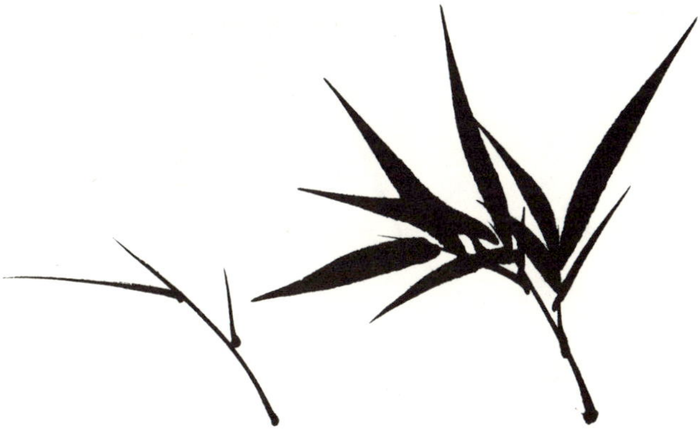

for left side of the stem

Note: Later leaves will be added to the end of the segments. This arrangement will suggest morning bamboo or bamboo in the sun.

2. Fish Bone has a middle segment, one on the left side, and one on the right side. But the left and right side segments branch off at a common point.

fish bone

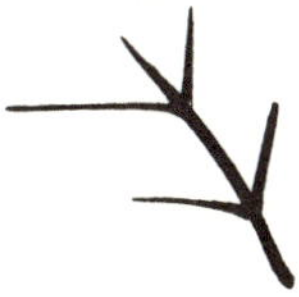

feather

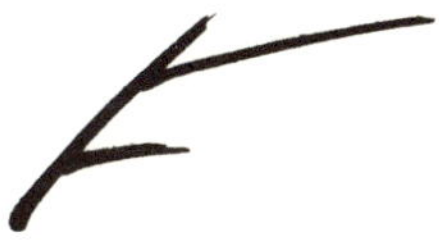

3. Feather has two segments from the same side of the middle segment.

   Note: If all curves bend in one direction it suggests that the wind is blowing or it is raining and all the leaves are forced to go that way.

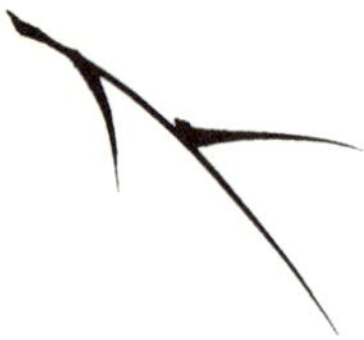

in the rain

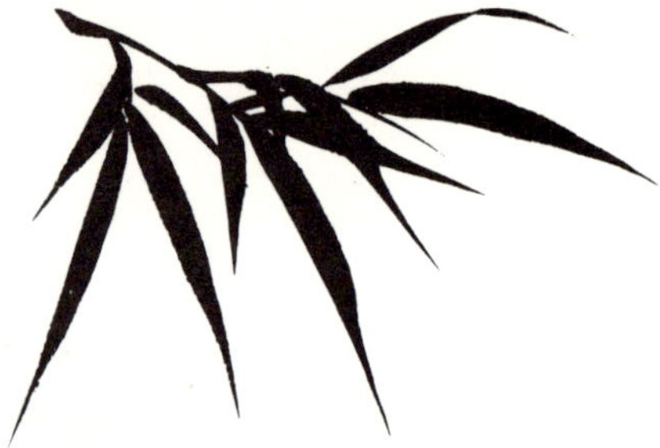

C. **Developing the Horn-of-Deer**

Continue to form the top of a tree:

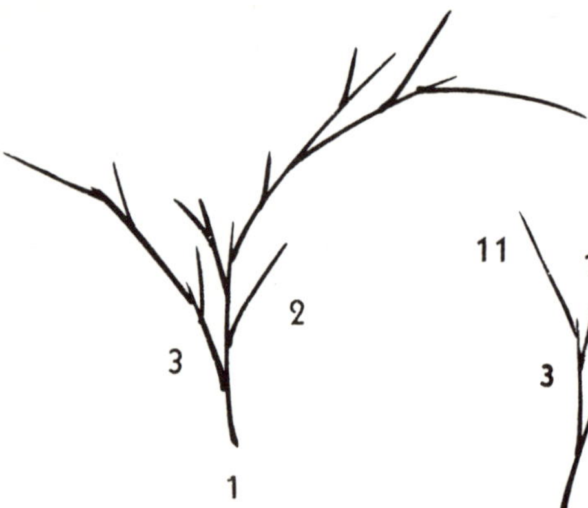

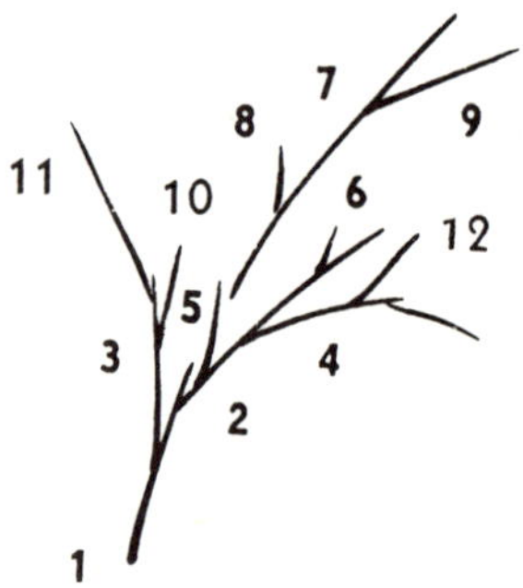

D. **Placement of Horn-of-deer on the Bamboo Stem**

The horn-of-deer grows from the joints of the bamboo stem, but only on one side of each joint. For example, if the first joint has a horn-of-deer growing on the right side, then the second joint will have a horn-of-deer growing on the left side. The horn-of-deer alternates. There are never two horns-of-deer growing opposite each other from the same joint.

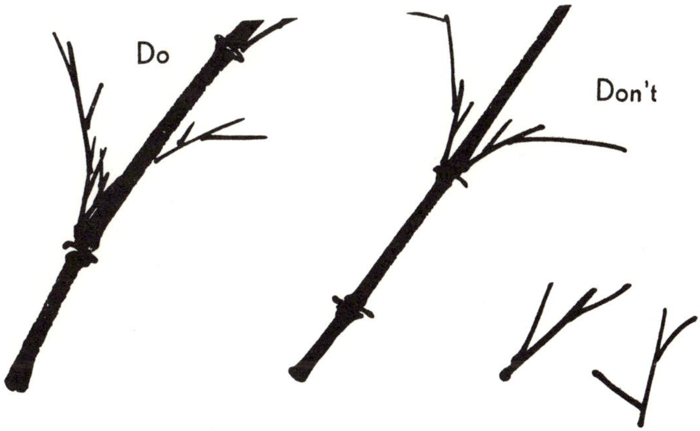

Note: Make horn-of-deer of different lengths so that some of the leaves will be closer to the stem and some will be further away.

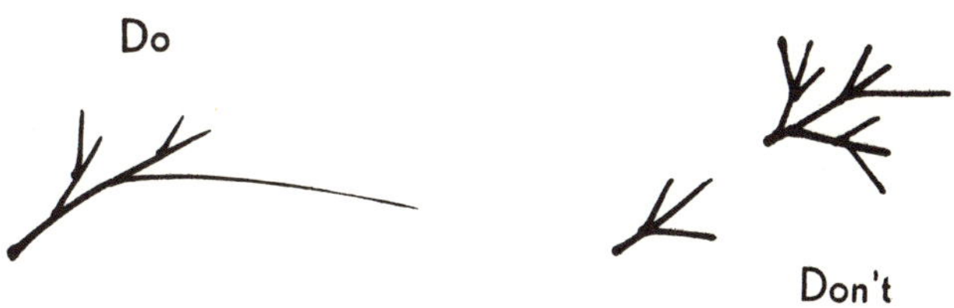

THE BAMBOO LEAF

A. The **stroke** for the bamboo leaf is similar to the stroke for the horn-of-deer, but there is no pause at the beginning and the line is wider in the middle. To make one stroke for one leaf:

1. Hold the brush upright. The tip of the brush is pointed.
2. Before the brush touches the paper the hand is in motion. The entire arm swings from the shoulder horizontally.
3. Without hesitation make the stroke in a smooth, slow flowing movement. Actually this is a straight line, but because the hand is swinging like a pendulum, it forms an arc and touches the paper with more pressure in the middle of the leaf while the beginning and end of the leaf are pointed.

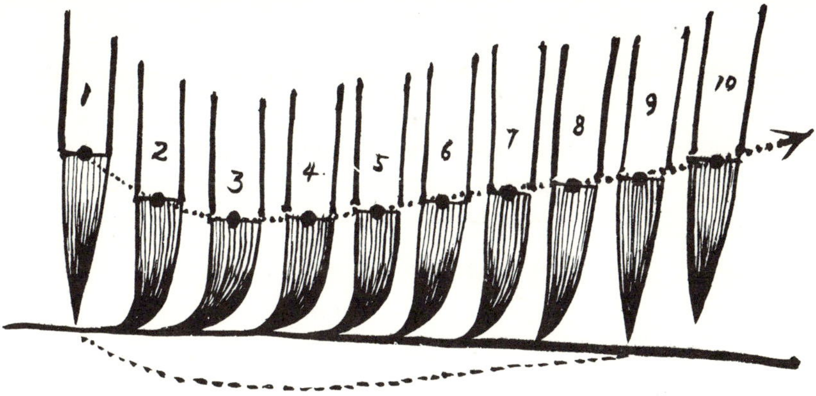

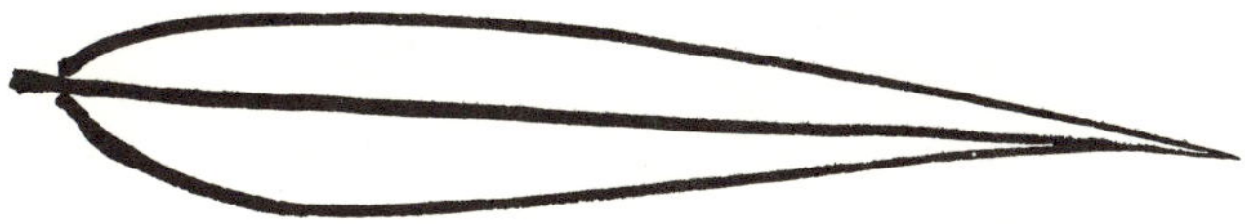

Note: Roll the brush slightly before making the next leaf. Do not move the body or swing from the wrist or the elbow. Avoid dropping the brush too suddenly or beating the paper too hard at the beginning of the stroke. If your leaf is wide at the beginning, you have dropped your arm too early; and if it is wider at the end, you have pressed your brush too much or lifted it too late.

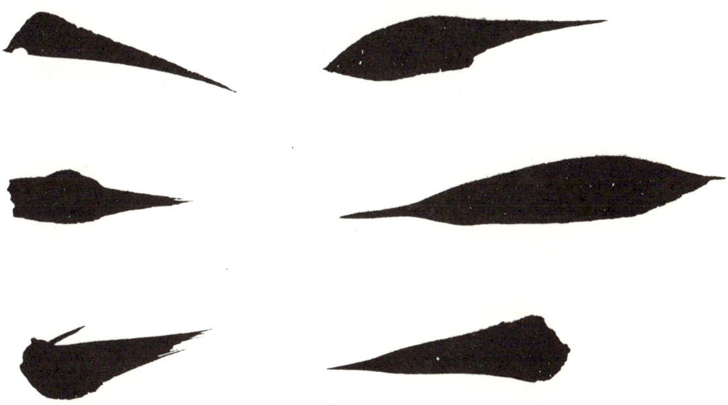

Don't press your brush too much or lift it too late!

If the end of the leaf is open or broken, you have moved too fast. The stroke will be made slower on newsprint than on rice paper. Keep the tip of the brush pointed and re-shape by re-inking or turning the brush.

B. **Ink & Color**

There should be dark and light leaves in a bamboo composition.

1. Make dark leaves first. (The exception to the rule is made when you paint on heavy card-like paper)
2. Make black strokes more slowly than green or gray ones. Black is stickier than green and gray.
3. For gray leaves the student may find it easier to make up gray ink in a separate dish by adding water to black to get the gradation desired. Remove excess moisture from the brush before taking up the gray.
4. A good combination for painting leaves is to take on the brush in this order: first yellow, then blue, then black on the tip.
5. Yellowish leaves should be near the root to show that they are older leaves.
6. Saturate the brush with ink to make the leaves.

C. **Kinds of Leaves**

Each kind of leaf comes from a segment of the horn-of-deer. However, not every horn-of-deer must have a leaf. Sometimes several different kinds of leaves are found in the same composition.

1. **New Moon** — The horn-of-deer has only **one** leaf. This is for a very new leaf or for a very old leaf where the others have fallen away.

2. **Fish Tail** —There are always two leaves on one segment to form the fish tail. They are on opposite sides of the segment and they do not start at exactly the same point, and they do not have to be of the same size. The fish tail usually shows the full front view. For a side view make one leaf long and one leaf short.

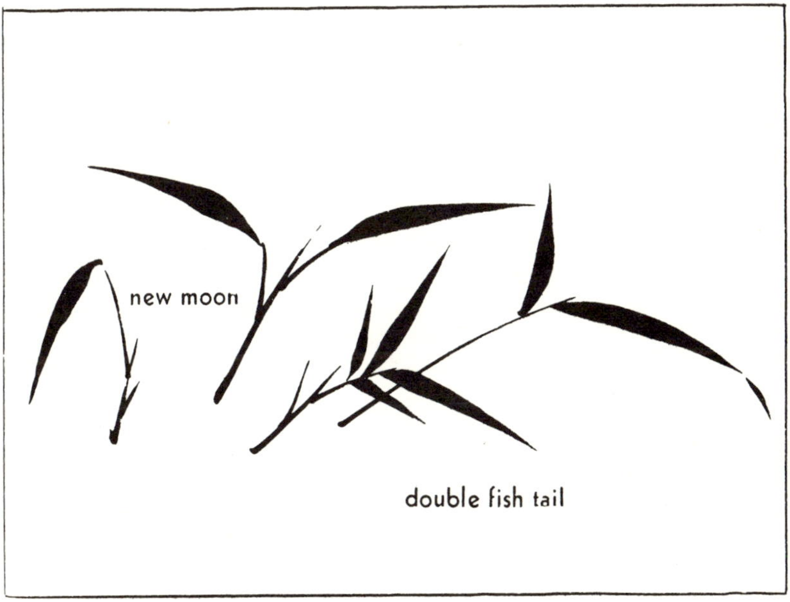

3. **Goldfish Tail**—There are three leaves on one segment to form the goldfish tail . . .one in the middle, one on the left side, and one on the right side. They do not come from exactly the same point. They are slightly separated and there is no overlapping at the starting point. The middle leaf is slightly longer than the side leaves. Follow the horn.

gold fish tail

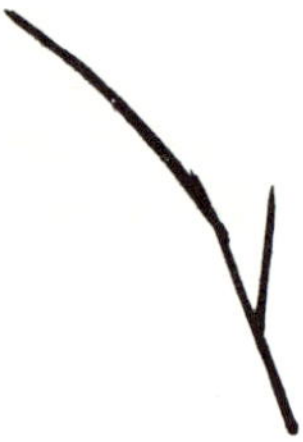

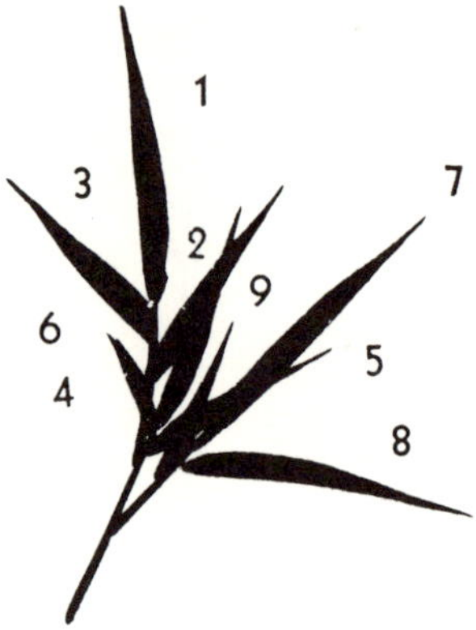

4. **Between** — Four leaves on one segment make up the Between. This shows completely the front view of four leaves. These leaves can be growing up, hanging down, or slightly to the side. The two end leaves on opposite sides of the segment are shorter. Do not overlap the leaves at the starting points.
Leaf number 1 and number 2 are longest
Leaf number 3 is short
Leaf number 4 is shortest

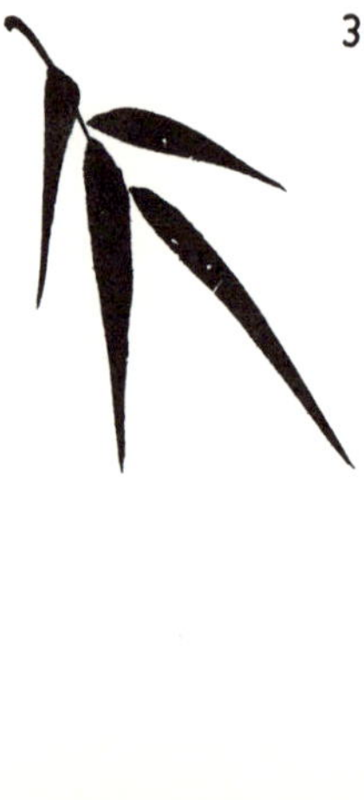

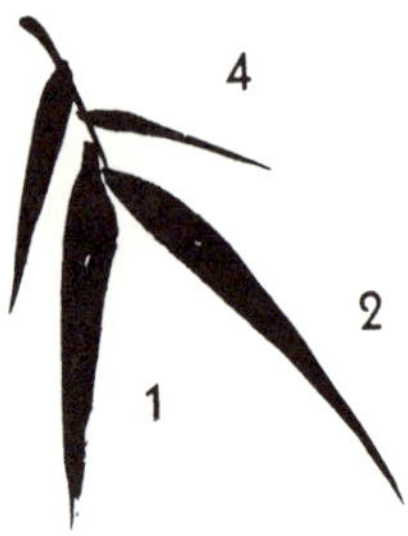

5. **Division** — Division is the side view of the four leaves known as Between. Three large leaves are on one side of the segment in three different positions; one is facing toward you, one turns slightly to the right or left. The fourth leaf is smaller and on the opposite side of the segment. Sometimes this fourth leaf may be omitted.

Note: Avoid making leaves parallel.

between

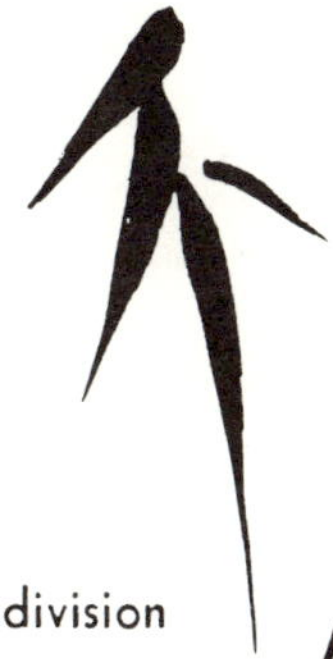

division

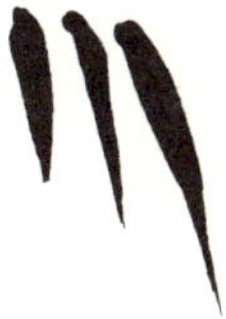

leaves parallel

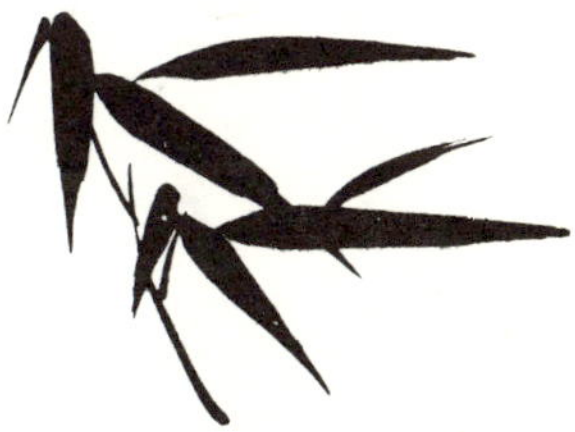

other style of "division"

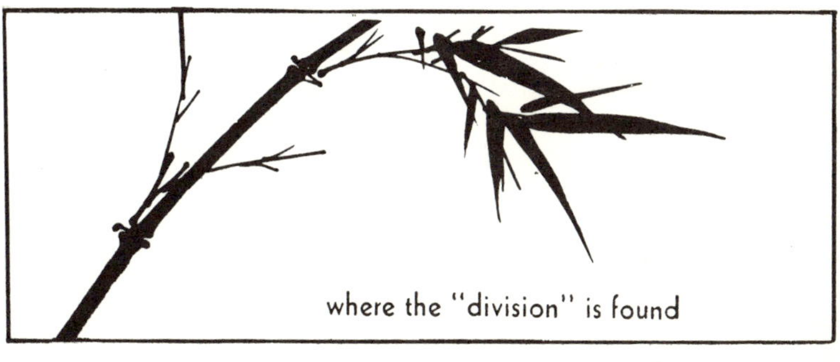

where the "division" is found

6. **Swallow** — The swallow is a bird's eye view of the Between. You are looking down at a four-petalled flower. This is used in front of the trunk or close to the trunk and on a short horn-of-deer.

There are five leaves for the swallow (sometimes Leaf 4 can be omitted)

Leaf 1 is a small leaf (it is the head of the bird)

Leaf 2 and leaf 3 are similar to the two leaves of the goldfish tail going downward.

Leaf 4 and Leaf 5 are larger leaves going outward like the bird is going to fly. Before making leaf 4 and 5 make a short line and then continue with the leaf.

wings of a "bird"

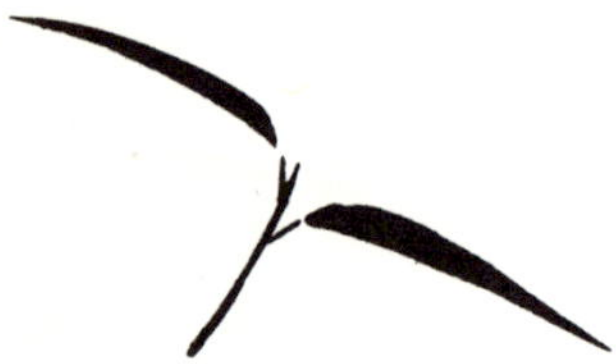

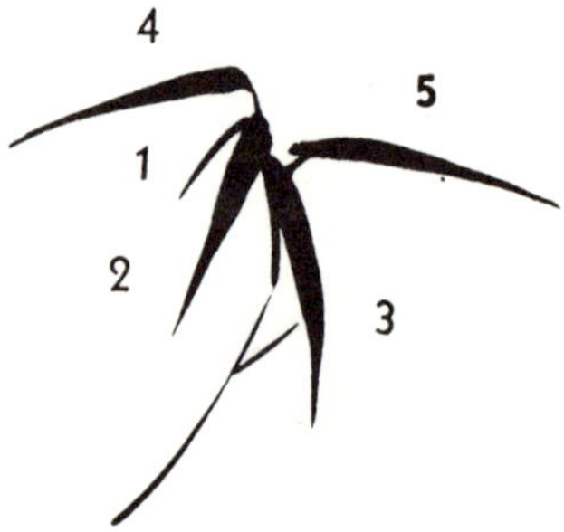

horn-of-deer going down

horn-of-deer going up

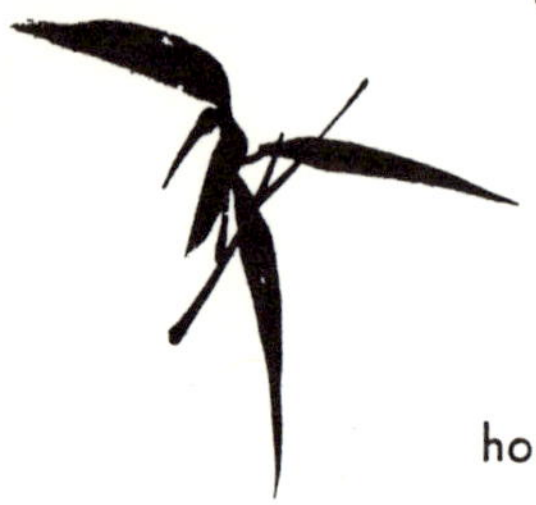

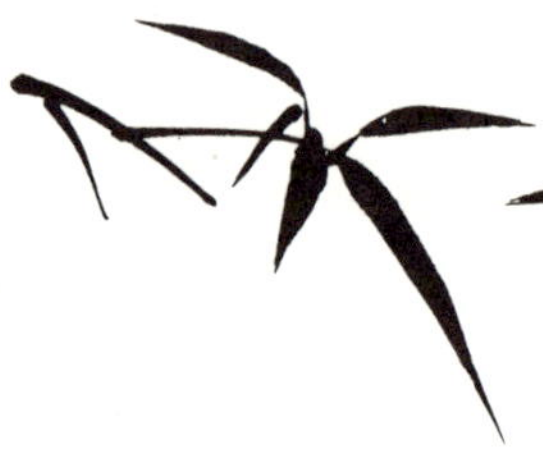

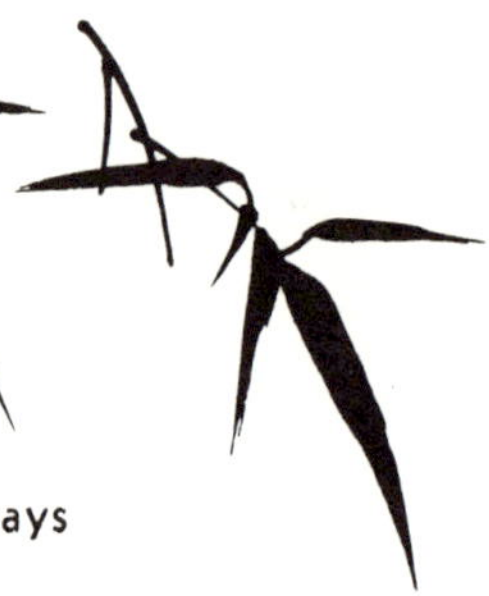

horn-of-deer going sideways

Note: Do not make leaves 4 and 5 too "up" or the bird will look too startled.

The swallow is always like this whether the horn-of-deer goes up or goes down. It does not matter which leaves are long, or which are short. Sometimes you stop short because of the position of the stem or trunk of the bamboo.

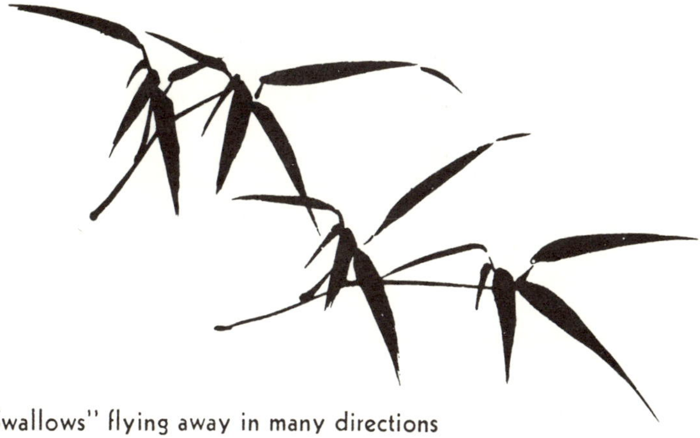

"Swallows" flying away in many directions

Some leaves shortened or even omitted

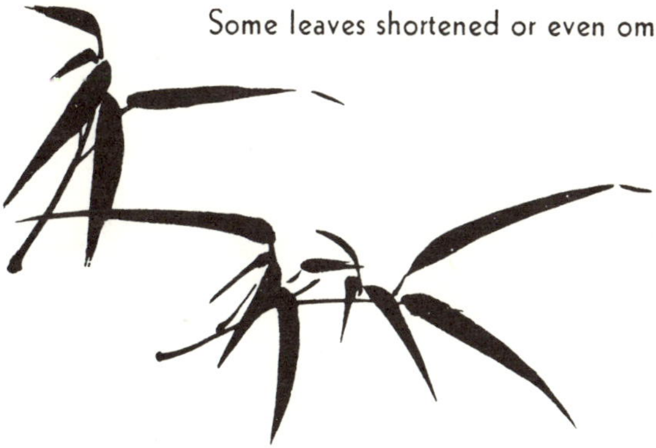

7. **Landing bird** is made with four leaves on a short horn-of-deer near the trunk.

Leaf 1 is a small leaf for the head.

Leaf 2 and 3 are for the body downward.

Leaf 4 and leaf 5 are two leaves going up but almost horizontal. Also make a short line before continuing these leaves.

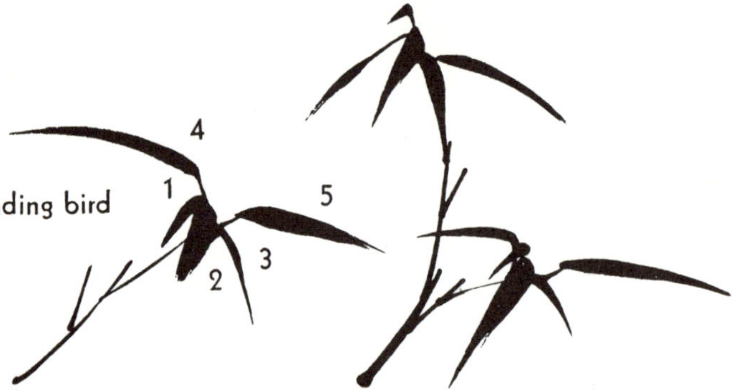

a landing bird

8. **Frightened bird** is made with five leaves on a short horn-of-deer near the trunk.

Leaf 1 is small and pointing upward.

Leaf 2 and 3 are almost horizontal or slightly downward.

Leaf 4 and leaf 5 are small and close to the middle segment of the horn.

Note: This bird is taking off.

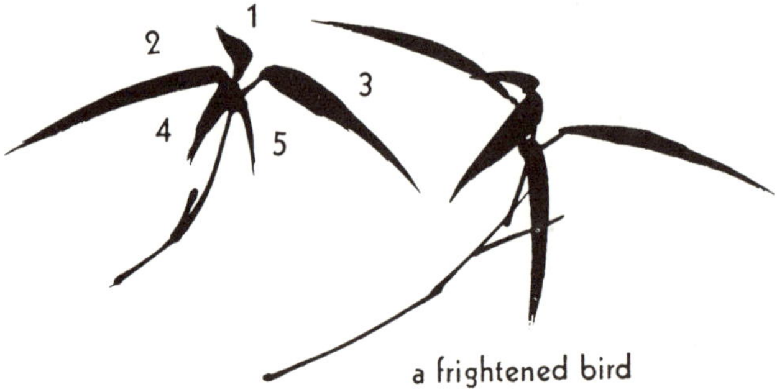

a frightened bird

Part III

GROUPING OF STEMS, BRANCHES AND LEAVES

COMPOSITION WITH MORE THAN ONE STEM

Here we are going to study the second portion of exercises on bamboo painting. In Part II, we learned the basic formations of stem, knots, branches and leaves, and dealt with each of them as a sole pattern. Now, we will study how to paint more than one stem together for the primary composition of a bamboo painting, how to group branches for those stems and lastly, how to put clusters of leaves which are all related to them.

"The first principle of a bamboo composition is, the four parts of the plant should be considered in the following order: stem, knot, branches and leaves. If this basic rule is not followed, time and effort will be wasted and the picture will never be completed." This is the beginning of the early Book of Bamboo, a part of the Mustard Seed Garden manual of painting which was prepared and published by Chinese master in the year 1701 A.D.

Following are principles in grouping stems. Before going any further, the making of sole pattern of bamboo stem must be reviewed. The Book of Bamboo says: "If you can not paint a single stem yet, what is the use to study how to paint two in one painting?" Please practice more on bamboo stem-making.

A. Avoid the same length of the joints for adjacent trunks and never make the trunks parallel to each other.

For the exercises on stem-making, (see Part II, page 42) Chinese art teachers always ask the students to put trunks of stem one after the other, yet with one thing to keep in mind, do not make adjacent trunks of stem with joints of the same height. It is also important to avoid parallel lines in making trunks and stems.

I should like to call your attention to the relation between this dislike of symmetry and the sentiment toward nature. It might be asked whether there is anything really symmetrical in nature. The branches of trees grow in the greatest freedom and irregularity; the stars of heaven are scattered with the utmost variety. It is this aspect of nature that inspired the Oriental artists in both their life and art. On the other hand, by the representation of motion in art works, Oriental artists manifest their intimacy with nature also. In sympathy with nature's vivid movements, we know the need of the spontaneity and irregularity of brush treatment. Such recognition come only from the subjective attitude towards nature; and such treatment come only from the keenest observation of nature.

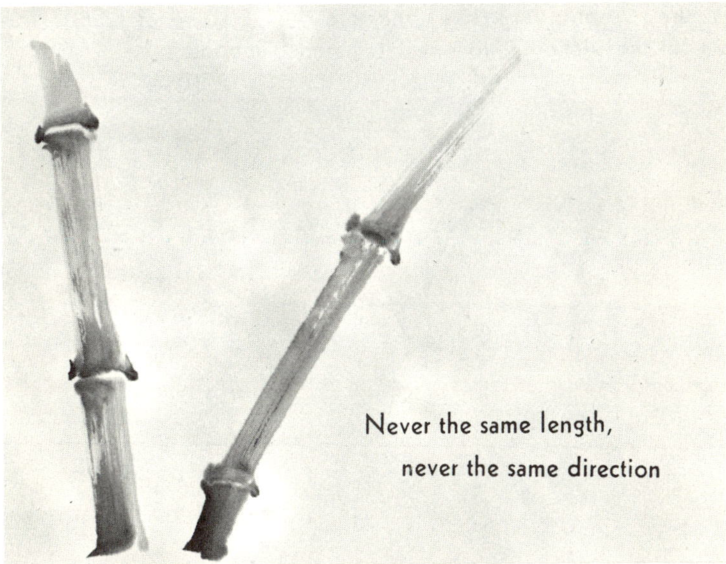

Never the same length,
never the same direction

B. Trunks, stems and branches should bend beneath their load.

1. Bamboo has springy stems. Since all bamboo has younger and weaker stems near the top of the tree, we discover that the thinner the stem is, the more it bends. So, we have to show more or less the tendency of bending. For a big bamboo, even it is partly shown in a small painting, the bending of a stem is unavoidable.

2. A bending stem is made with a circling movement of arm and brush. Think of your shoulder as a fixed point, which is the center of a big circle, and the distance between the brush and your shoulder remains always the same as the radius. By waving of your arm together with your brush, which is held by your hand upright to the paper, you always can have a long curve line, which is only one portion of a large circle, (a segment of circle.) If your paper is rather small, try to have your elbow as the center of circle. You may rest it on the table, and make shorter curve lines on paper by using the same way as before. Now, instead of making one smooth and continuous curve line, try this time a bamboo trunk which you have learned before. From tip to root or vice versa, although the stem are painted section by section, the continuity of the idea expressed through the brush remains intact. The Book of Bamboo says: they should look like "the thread of a string of coins," to string all the sections into one integrated bamboo.

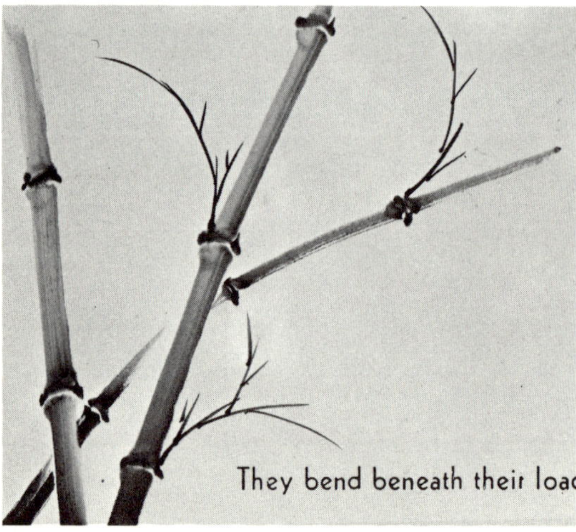

They bend beneath their loads

C. Variation of size and shade for a composition with two stems.

How can flatness in arrangement be avoided? Or how is depth in the painting suggested? The basic rule is: If we can not have the two in different colors, we must make them in different sizes. We usually use the lighter color for the one supposed to be at the back or in the distance.

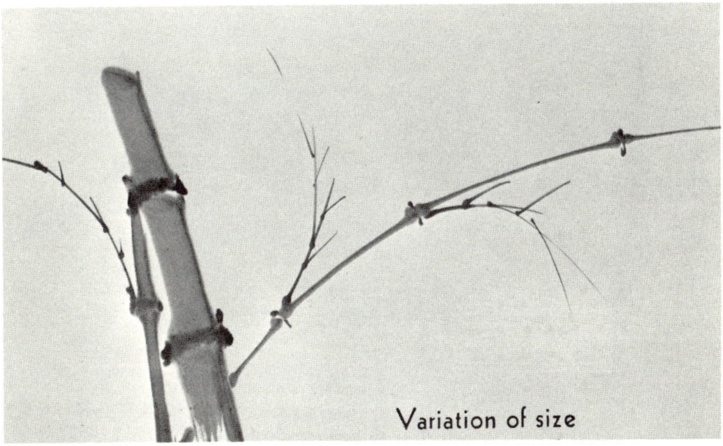

Variation of size

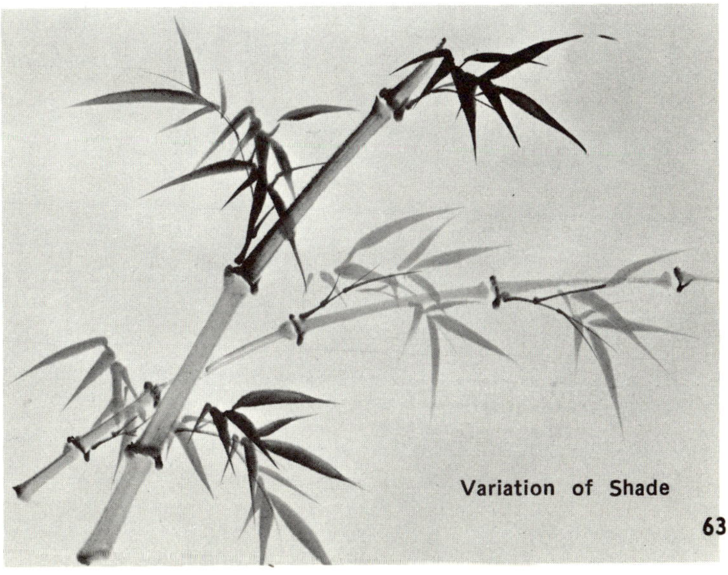

Variation of Shade

D. More treatments for a composition with two stems.

The easiest way to create the feeling of three dimension in a linear design is keeping the arrangement and even the patterns themselves related or associated to every variety of triangles.

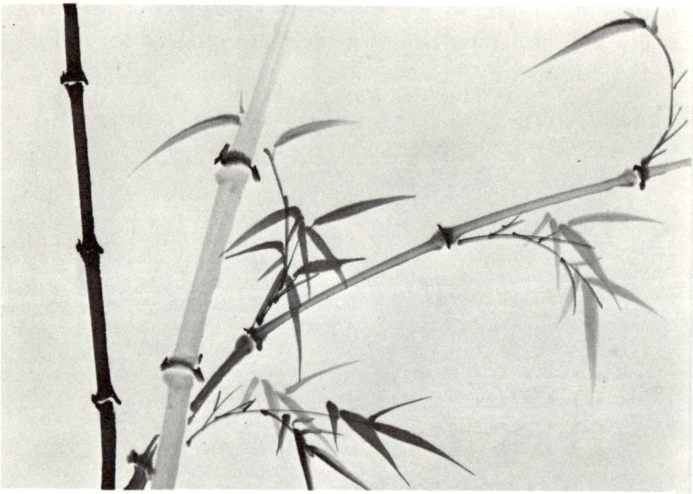

Strongly suggests the top is moving forward and approaching us, or two trunks are coming from the same root.

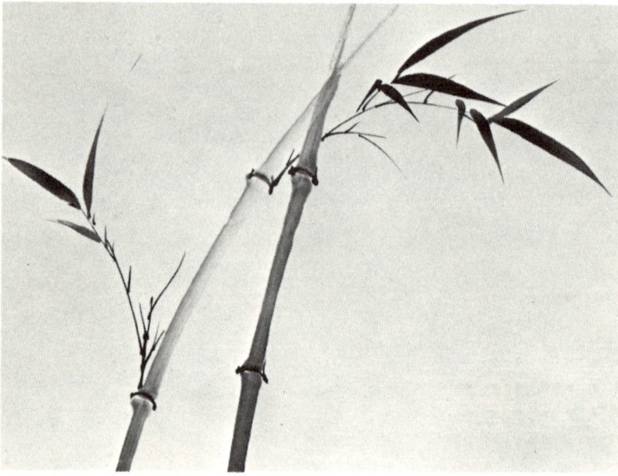

Strongly suggests the top of the tree is either higher than our eye-level, or is crossing the other.

E. The way of making a stem behind another.

To make stems cross each other with solid black will not show the space between them. It is not convincing to tell they are separated afar or one of them is behind — even when diverse color or varied size is applied.

The overlapping of heavy black strokes makes the rice paper too damp and weak. Lighter color will bring uncontrolled running of ink at the crossing point.

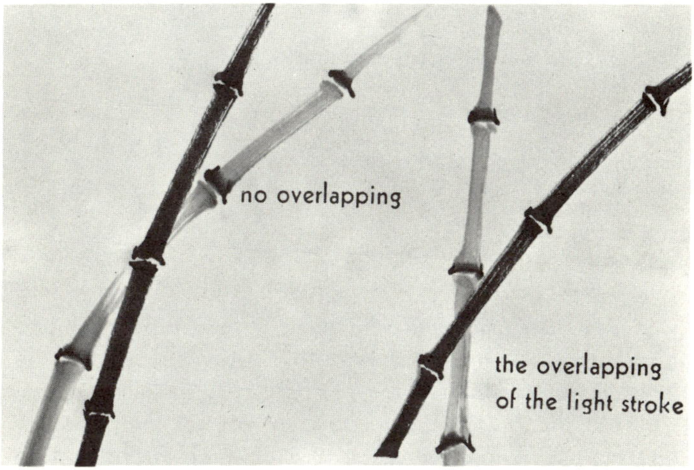

We must make the one in front first. It is completely independent and always in strong color. For the one passing behind, we usually start our strokes section by section from the tip to the root. Before the brush moves too close to the stem which is planned to be the front one, we pick up the brush gradually and then in the same manner, drop the brush after it passed the other side of the front one, and continue the movement of "stem-making" to finish the second bamboo we want.

With regards to the movement of the second stem, actually there should be no interruption. Every movement makes up a complete bamboo stem. The only difference is: this bamboo-making movement has certain parts accomplished in the air, or rather away from the paper, while the brush is crossing the first bamboo.

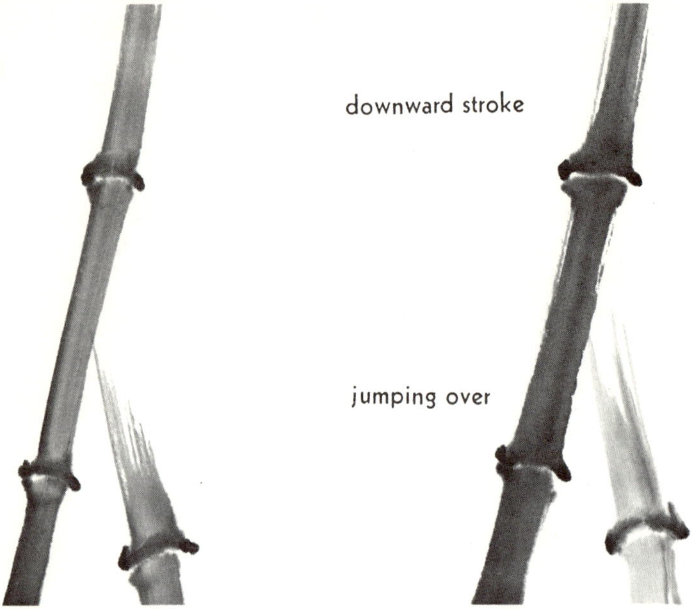

Please pay no attention to which part of the second bamboo is going to be made in the air or above the paper and whether the end of one section or a joint is going to fade away partially or completely behind the first one. In short, except for the "jumping over," the instruction in stem-making in Part II of this book (see page 36-42) should by all means be followed.

The technique of Hsia Chang.

Some famous artists of earlier period made no crossing trunks. Wen T'ung, the greatest Sung painter of bamboo in ink, made only one branch sweeping down in an "s" curve from upper left to the lower right. We know Hsia Chang (1388-1470) is the first artist who made bamboo bordered stream in the Spring Rain, (1441, the Art Institute of Chicago Collection.) with light gray stem in front of dark ones, and one heavy gray stem covering a very pale trunk. The one which crosses the other at the back suggests the depth very effectively. **The proper procedures to be followed are as follows:**

a. Paint the lightly shaded stem first.

b. Finish every detail of this stem: including joints, knots and horns-of-deers.

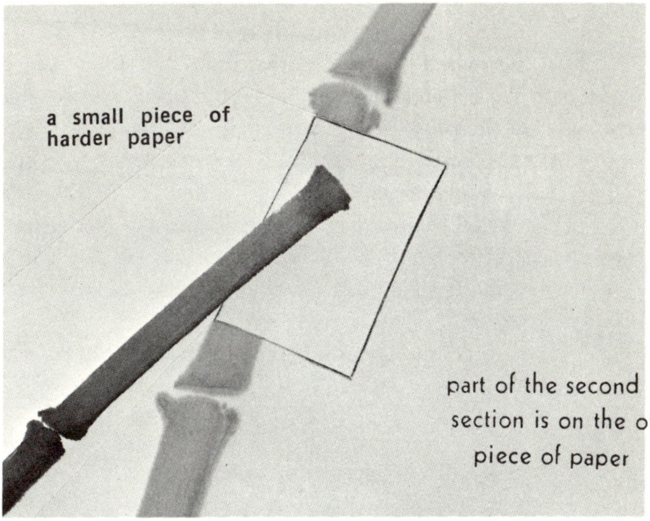

c. Prepare a small piece of harder paper with the width big enough to cover the width of the lightly shaded stem. This small piece of paper must have straight and clear-cut edges on both sides.

d. When the light stem is nearly or totally dry, use one edge of this small piece of paper to cover the very portion of the light stem where your black brush for the second stem is going to cross. Make the second stem as you want, but stop the procedure when your brush runs over the small piece of paper which is now still lying along the edge of the gray stem and make sure this paper is not pushed by the stroke of your black brush. Your left hand will be used to give pressure on it.

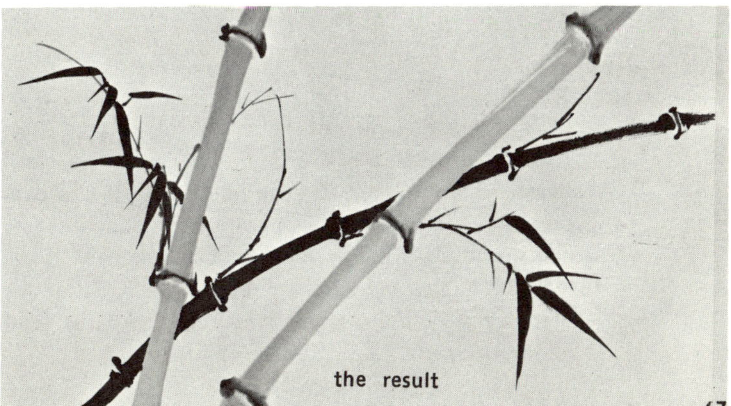

e. Start the rest of the black stem from the other side of the light one with the other edge of that same small paper again to prevent any overlapping or overspreading of ink.

Caution: (1) The small paper should be thick enough to stop the penetration of ink.

(2) When the edge of the small piece of paper becomes too damp because of excessive ink, it will warp after the ink dries and the paper cannot be used again.

(3) The continuity of the second stem should be watched carefully.

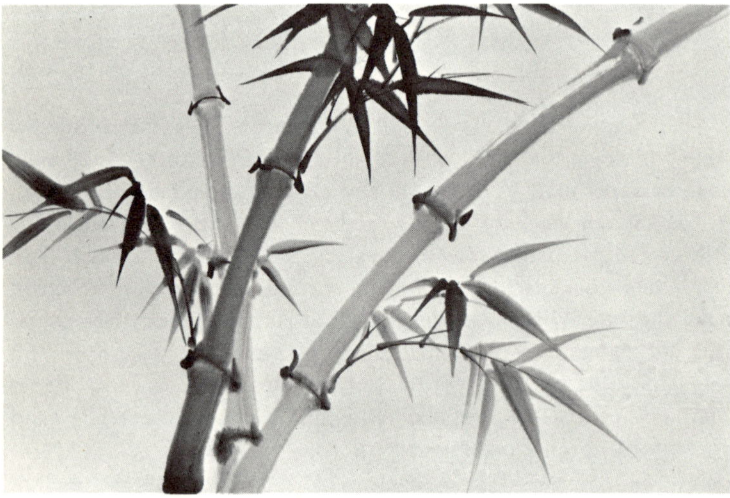

A complete painting

How do we make a light colored stem across a dark or black one? The method is much simpler than the above. The light one just runs over the black or dark one without stopping or jumping and the result is the appearance that it is passing behind. But the light stem must be made after the black or dark stem has dried already and the light ink must be clear and free of opaque color.

F. The arrangements of three trunks for a bamboo composition.

How is a pyramid drawn with the use of three to five lines? What are the different ways in constructing these lines?

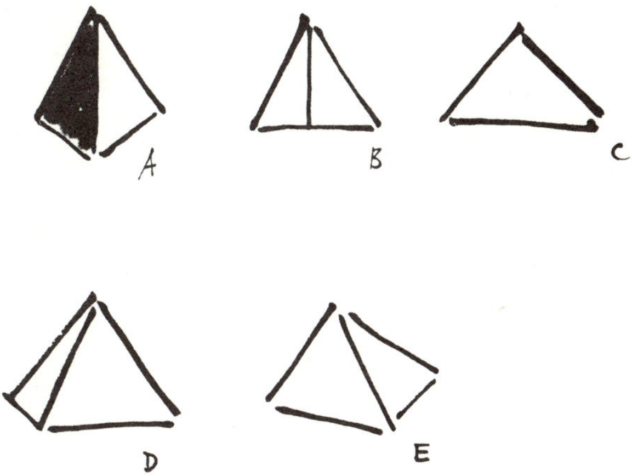

Pyramids A and B can be made with one straight line dividing the top into two equal parts. They are poor methods in drawing a pyramid. Without shading one side of the pyramid, the lines create only an illusion of flatness or a frame with three steel bars. Pyramid C is out of consideration as it shows only one side and has no suggestion of depth. It is very clear that we prefer pyramids D and E since solidity or volume is definitely represented.

So when we put three bamboo trunks in a painting, an upside down pyramid should be the guide for the arrangement. This means that the three trunks must point to different directions, spreading in the radiation. Never let the angles formed by the trunks be equal and don't make these trunks appear tied up like a bundle of flowers. Chinese masters call this a "drumstand," one of the many defects in stem arrangement.

Since no parallel lines are allowed, in placing three trunks in a single composition, we can have only two forms of arrangement: inverted pyramid D and E.

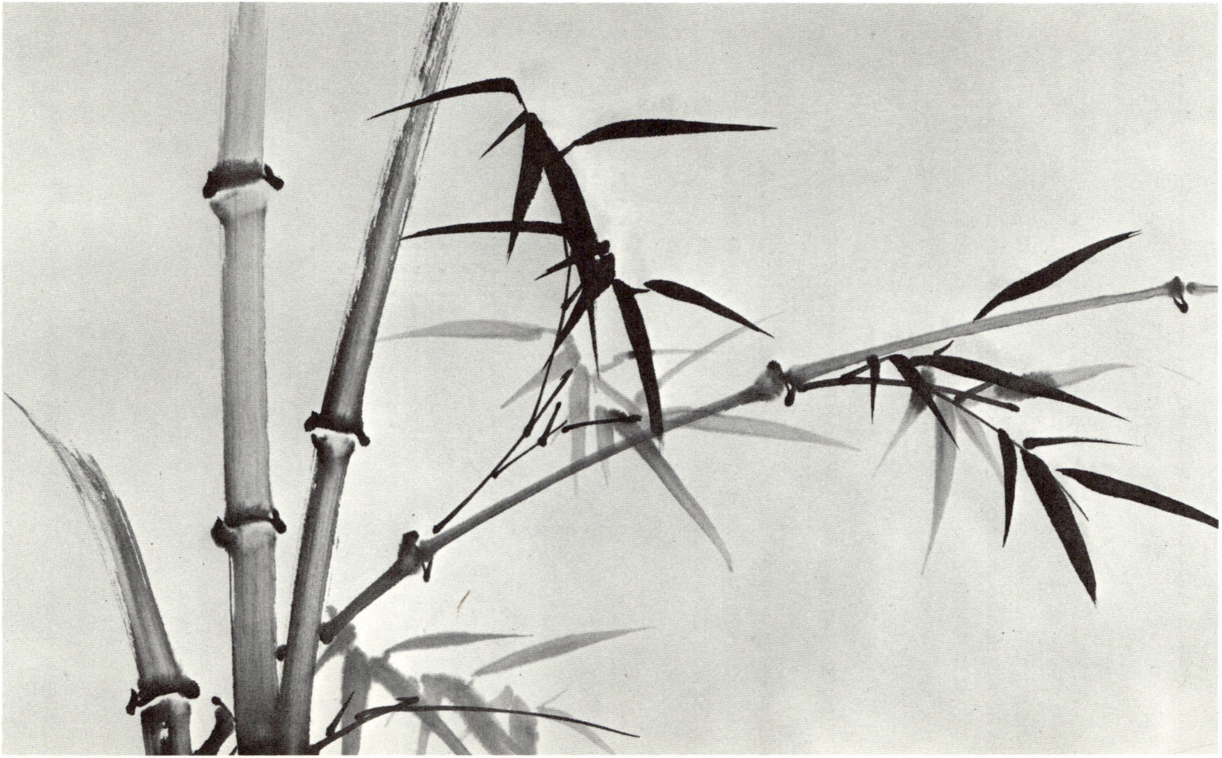

Yellow bamboo
(an upside-down "pyramid")

by I-Hsiung Ju

When any two of the three happen to cross each other, we have two ways of arranging only. For Chinese numbering mark system they are "fourteen" and "forty one" but in Roman system, they are IX and XI.

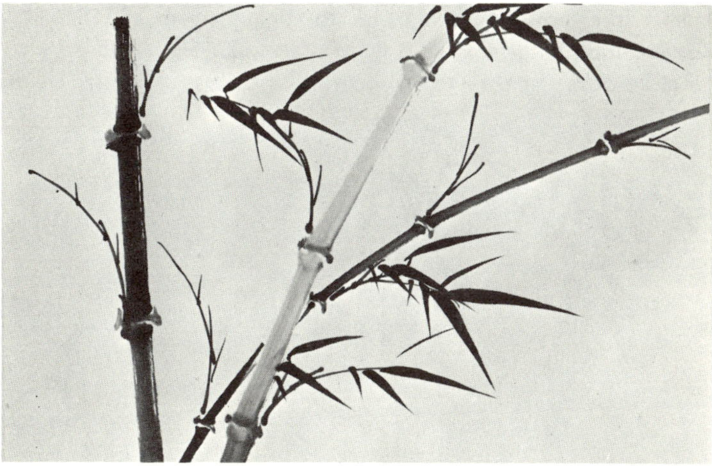

IX arrangement

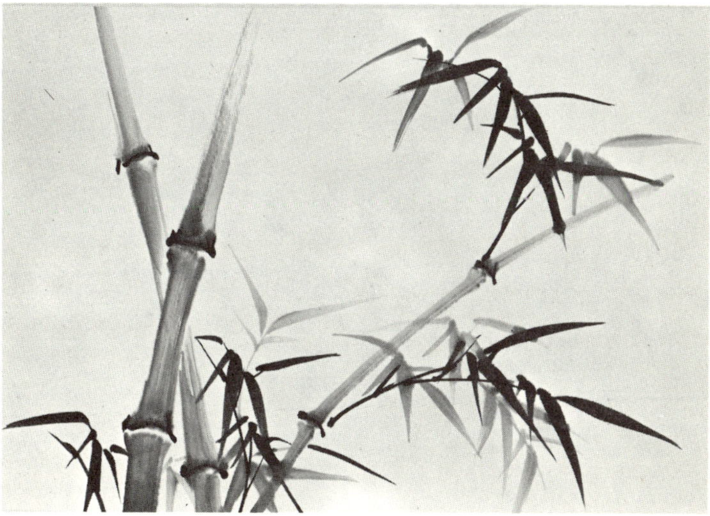

XI arrangement

Two trunks in front and a third one crossing behind is also permissible. But the gradation of the bending, variation of the size of angles and the height of the crossing points should be carefully studied. While bigger trunks have lesser tendency of bending and thin ones more, the division of the whole compsosition by trunks is a most interesting task for beginners to experiment on. Composition with more than three trunks as well as those.with two or more pairs of crossing bamboo will be discussed in later chapters.

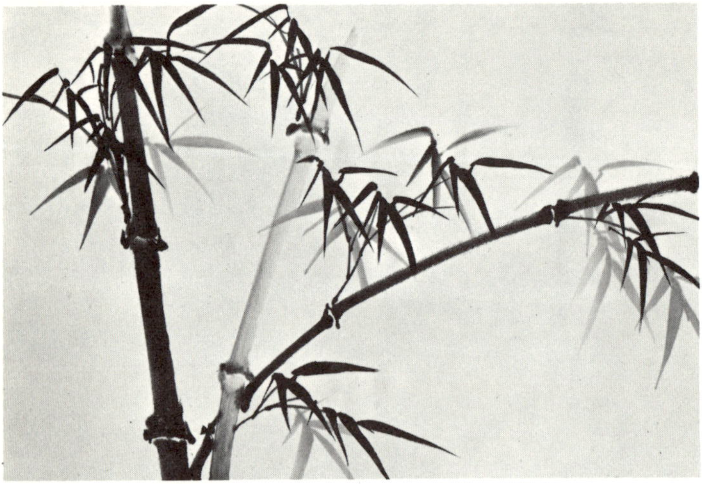

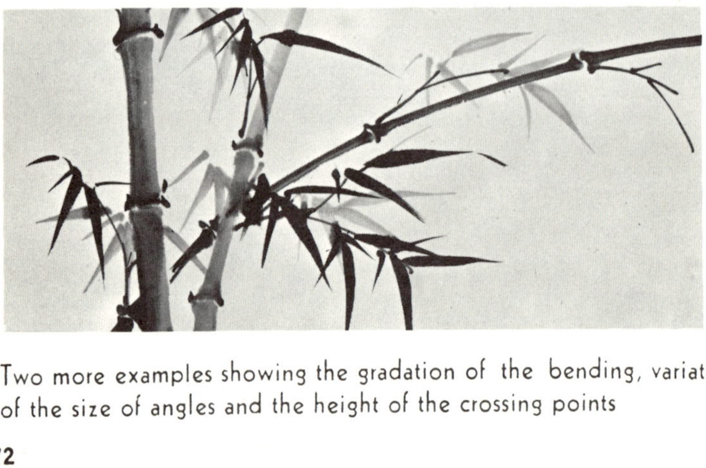

Two more examples showing the gradation of the bending, variation of the size of angles and the height of the crossing points

CLUSTER OF LEAVES

A. **The relationship between "horn" and "leaf formation"**

In Part II, we have the practice on leaves. Those that look like "fish tail" and "swallow" are called leaf formations.

The master of bamboo painting in his Book of Bamboo, a part of the **Mustard Seed Garden** says, "The manner in which a leaf grows from a branch should be clearly defined. If the branches do not grow from the knots correctly, the leaves will be disorderly and without proper support. Even if the stem has been placed beautifully and bending gracefully, when the branches and leaves are exaggerated and poorly composed and the direction of the leaves are not systematically related, the whole bamboo in your composition will resemble a flat board with thorns sticking out."

What then is the manner in which the leaves grow? How do the branches shoot out from the knots correctly? And is there any way to simplify the relationship between "horn" and leaf formation?

First, let us study the "horn." Whether the horns-of-the-deer are standing upright, stretching sideways or hanging upside down; the branches spreading out on both sides together with the central branch or major stem always follow the principles of alternating and progressing arrangements.

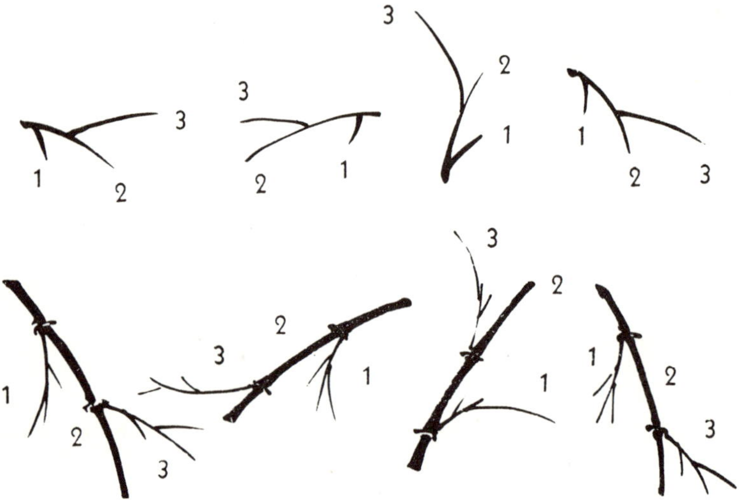

For instance, if we take any one figure of the above, we see foreshortened branches. Branches #1 are stretching towards the viewer and bending accordingly with the weight of the leaves growing out of their stems. Branches #3 are pointing sideways showing its full length with side-view to our face while branches #2 act as the middle ones and bisecting the angle formed by branches 1 and 3. But since we are viewing the whole "horn" from different directions according to the principle of perspective, there are no actual "equal angles" to be seen. And by the way, branch #3 may be placed very near the tip with a narrow angle. Sometimes, it can even be omitted to suggest that it is so close to branch #2 or nearly covered by it already.

Take a look at Fig. A. When the "horn" with #3 is that low, it does not always mean that this "horn" has been squeezed tight. This is the proper way to depict a "horn" at our eye-level. The "horn" now is lying flat with its side-view to our face. Then it is very clear that Fig. B is a "horn" also going sideways but at the position lower than our eye-level.

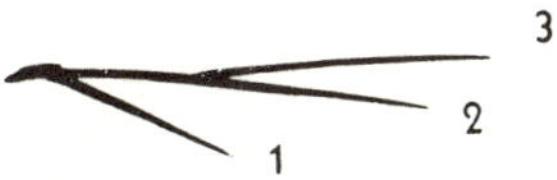

Fig. A

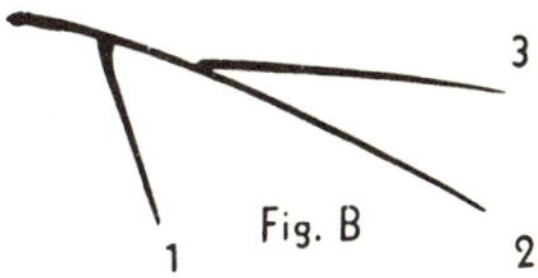

Fig. B

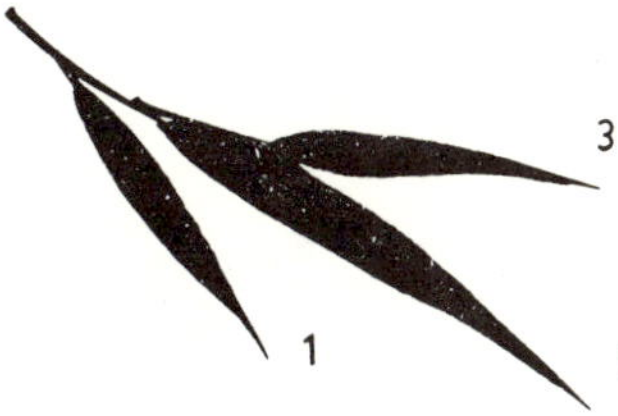

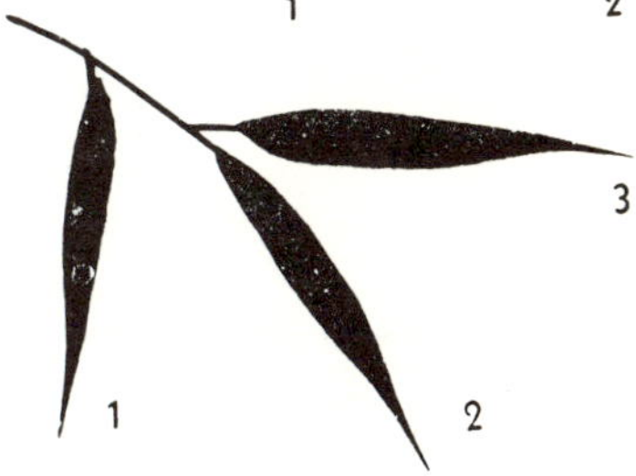

When the "horn" is vertically placed, it can be the front view or the back view of the same "horn"—Fig. C & D—and the side view of it—Fig. E & F—but never suggest it is either higher or lower than our eye-level.

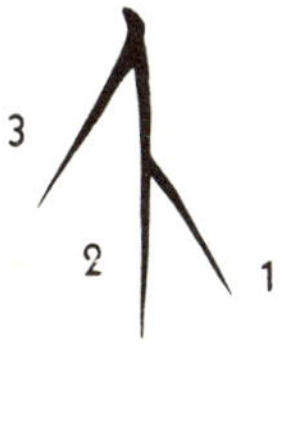

Fig. C

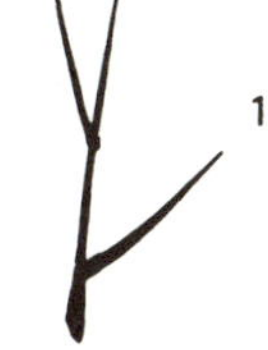

Fig. D.

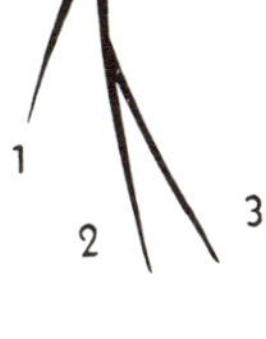

Fig. E

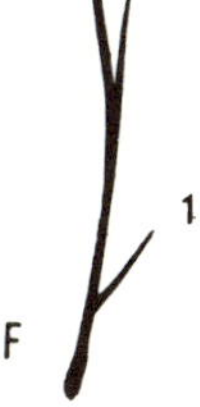

Fig.F

All branches shooting out from the knots follow the same principle mentioned above. Any transgression will make the leaves either too sparse or tangled in a mess.

Based on the same principle again, the students are required to practise the arrangement of leaf blades as shown below:

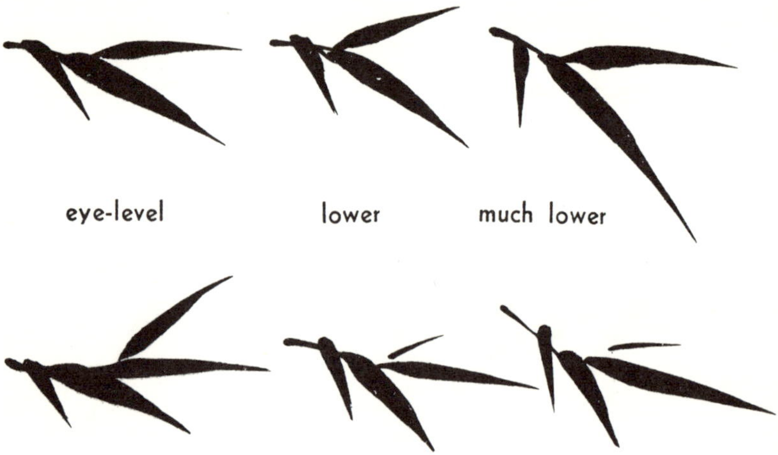

If we turn them to other direction, we will have:

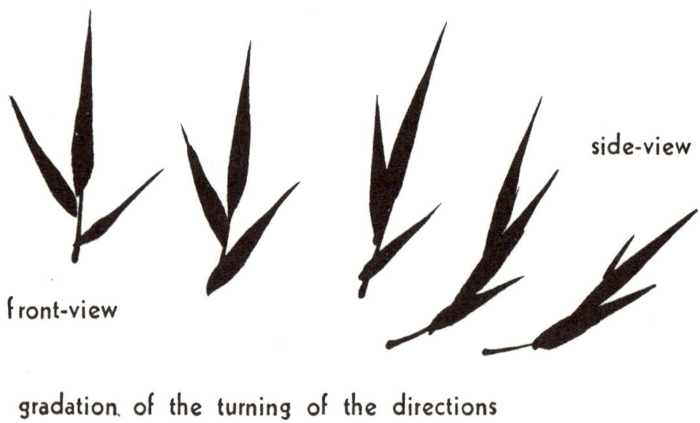

gradation of the turning of the directions

We still have several points to discuss about leaves:

1. **Young leaves**

Young leaves are not small leaves. The youngest leaf looks like a pin or needle. It is always found at the tip of a branch, at the end of the younger and smaller "horn," and easily mistaken for the extension of the branch. When the pin-like young leaf grows, it unrolls itself into the form of a sword with a broad blade.

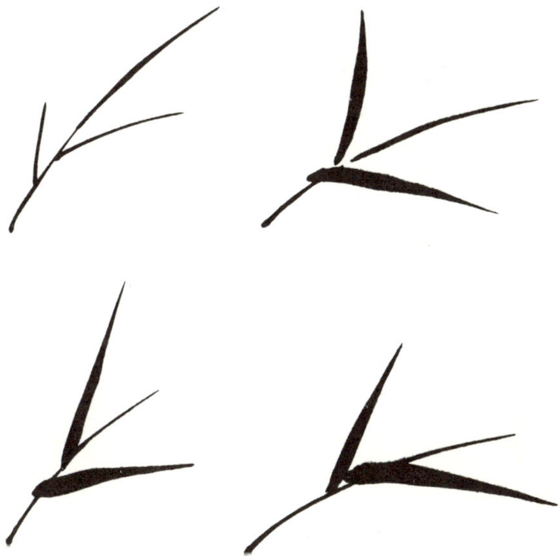

2. **The petiole**

It is not necessary to make a petiole for each bamboo leaf. Some of the leaves show no petiole, but the relationships between leaves and branches suggest very clearly the existence of every petiole. Before we make a turning leaf, or one which is spreading out, such as the "wing" of the "swallow," we always make the petiole first to depict the leaf as raised, or growing away from the branch.

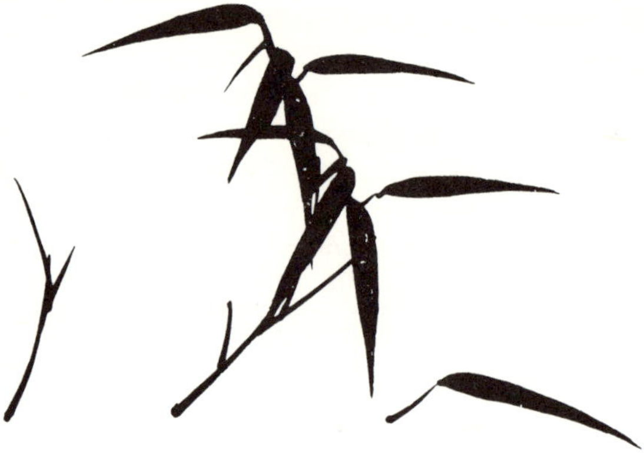

3. **The holder of "leaf-formations"**

Some leaves sprout out from the tip of the branch, some from the side. All these leaves belong to one branch but spread out palmately. In any kind of "leaf-formation" the branch itself bisects the whole aspect, and is the sole support holding each and every leaf of the "formation." By following this principle, all the leaves will be in good order and properly supported.

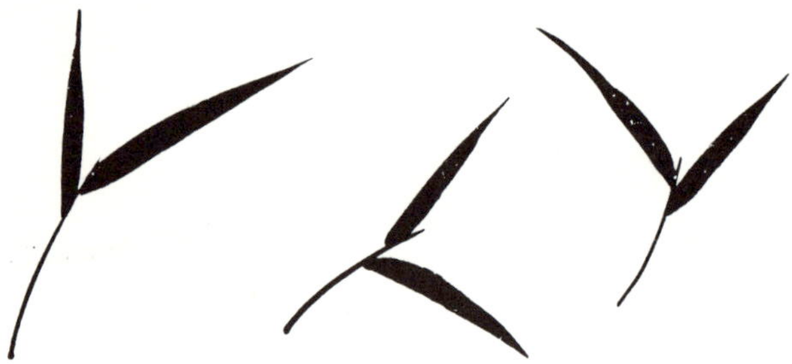

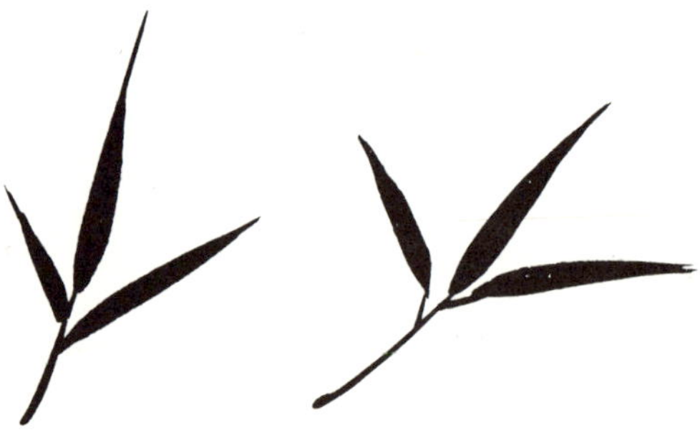

4. **The directions of diverse "leaf-formations"**

When we start to add leaves on "horns," we must remember that leaves need air and sunshine; so all the "leaf-formations" should show the tendency to grow outward and upward at any time. When branches bend with their loads, the leaves at the end of the branch will either point horizontally to the side or hang perpendicularly.

The different "leaf-formations" are actually different views of similar clusters of leaves, some in side-views, some in front-views and some in top-views. We must note that these leaf clusters have

side-view

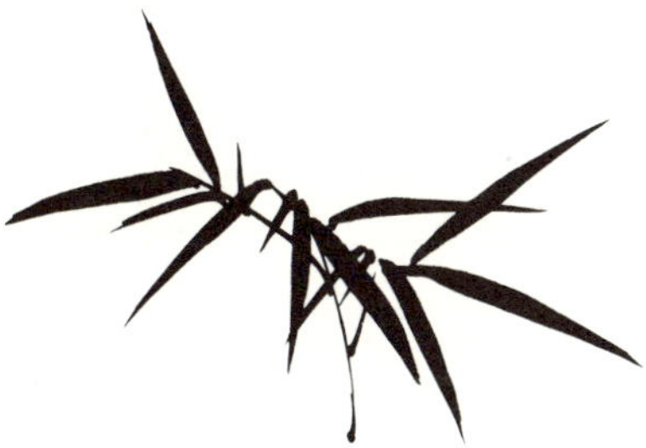

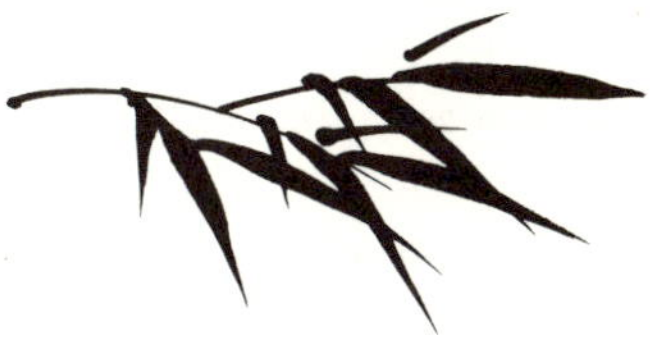

front-view

one important characteristic in common, namely, they always welcome more air and more sunshine with their out-stretched arms.

Only leaves belonging to the branches behind the trunk are seen in the back-view. Since the back-view leaves are in the distance, we make them in gray or lighter colors; and the supporting branches which are stretching out away from the viewer, should never be covered by these leaves.

This is what the **Mustard Seed Garden** means when it says: "The front and back of the cluster should be properly related."

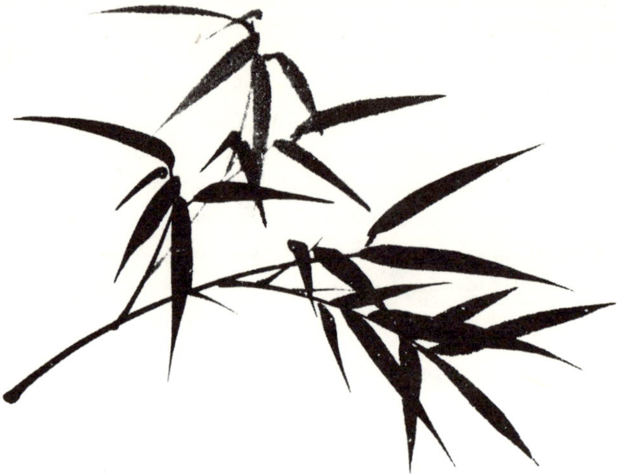

5. **The "dragonfly wings"**

The master said: "Avoid the 'dragonfly wings'!"

Leaves at the tip or near the tip of the branch are always younger ones. They grow closer to the branch and also closer to the other young ones. They sprout out briskly and erect. Old leaves came earlier, so they should be placed near the joints or bigger branches. They tend to fall away from the branch and look sparse and scattered. Therefore, at no instance are leaves in the form of the dragonfly wings.

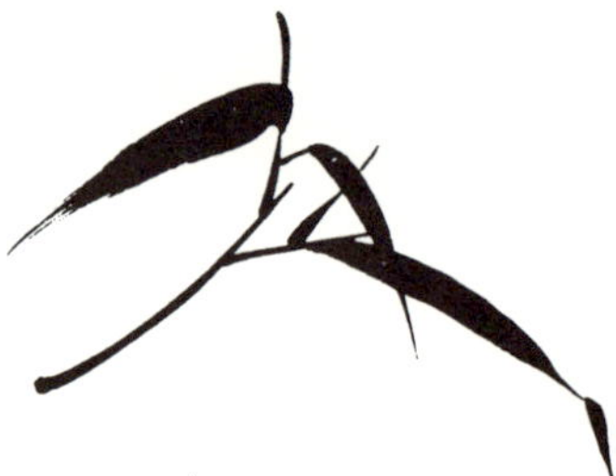

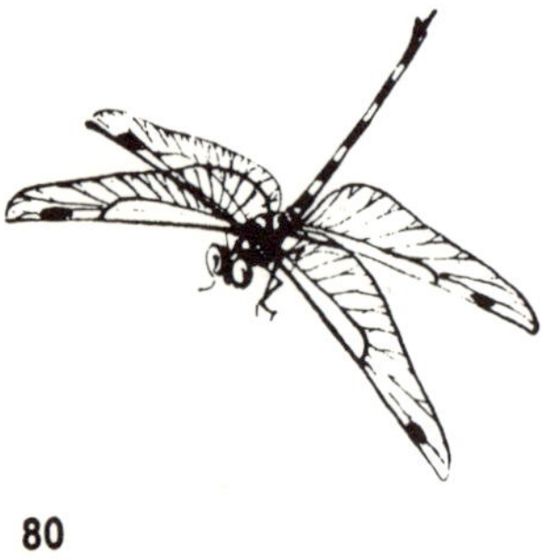

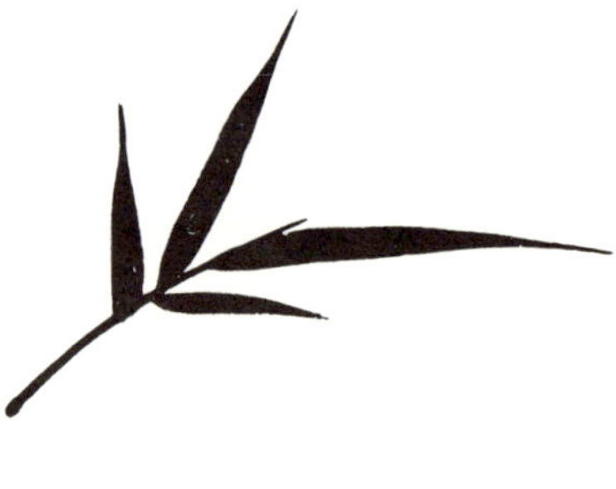

In short, bamboo leaves are never parallel to each other by nature. And since the "horns of the deer" suggest the directions of the branches in the first place, all the "leaf-formations"—which are systematically connected to the branches as I mentioned before—must follow the same directions. Since the position of each "formation" is different, even actual parallel leaves will show to the viewer the tendency of converging inward or spreading outward according to the principle of linear perspective.

So, we must avoid not only "dragonfly wings" but also formations that look like the Chinese characters **ch'uan** and **ching**.

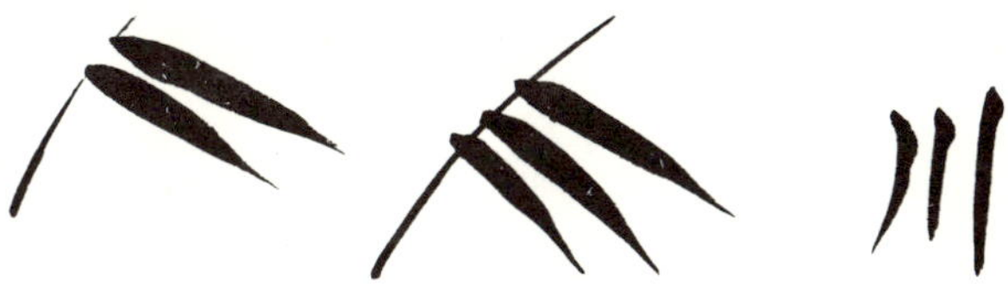

Ch'uan, the river

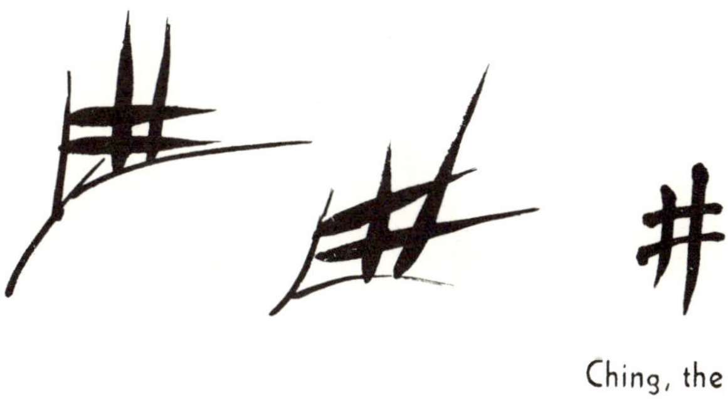

Ching, the well

B. **Repetition of the "leaf-formations"**

1. **Varied sizes and varied angles**

We have discussed the perspective of "horn" and "leaf-formations" earlier, but only for the single patterns.

To show the illusion of three dimensions, we use the principle of linear perspective. For example, with two leaves, in the "fish tail" formation, we always make one of them shorter, and the other one closer to the center or the stem. The "goldfish tail" follows the same rule. The shorter leaves are not really short leaves; they are only foreshortened. The distances between leaves and the variety of angles from which they are viewed should be prudently studied.

2. **Cluster of leaves by repeating the "leaf-formations"**

The repeating of "leaf-formations" has been mentioned before. We are going to have new exercises for practice here regarding the principle of linear perspective. Please take note of the gradation of direction of the "leaf-formations" and see how it helps to show the mass or thickness of each cluster. When one of these "formations" is misplaced, or the gradation of the turning is not followed, the cluster of leaves will look flat and shallow.

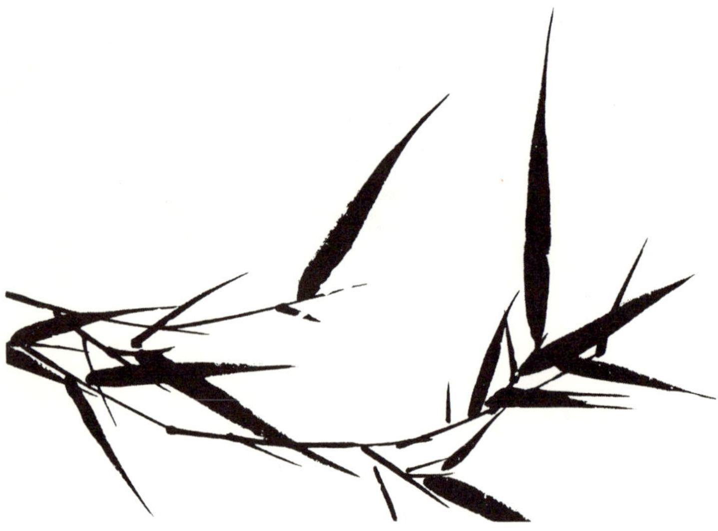

Cluster of repeated "fish tail"

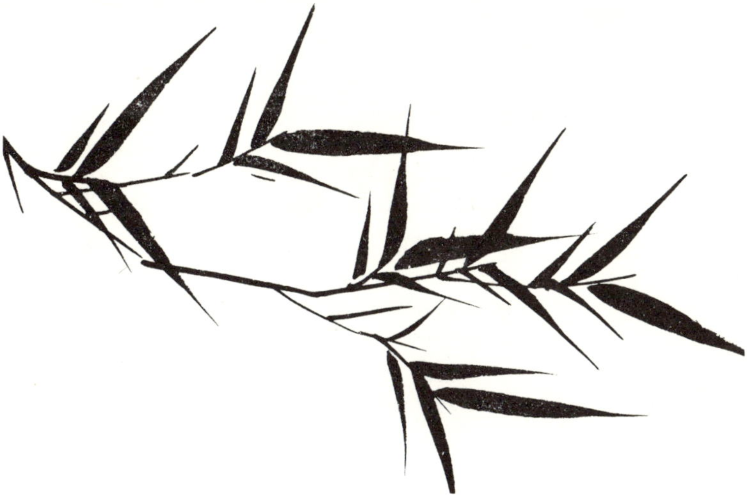

Cluster of repeated "double fish tail"

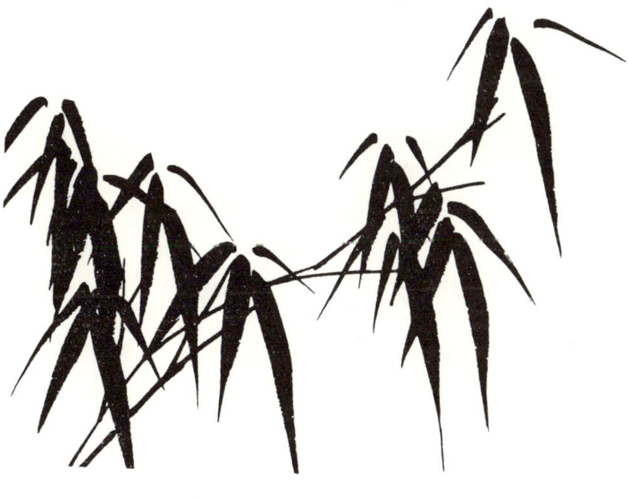

Cluster of repeated "between"

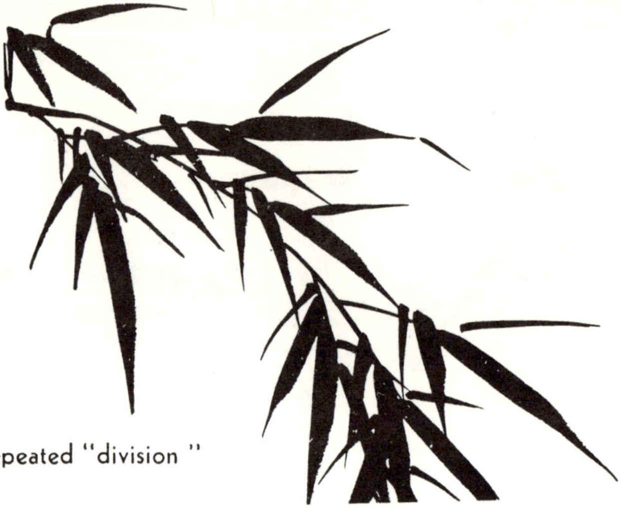

Cluster of repeated "division "

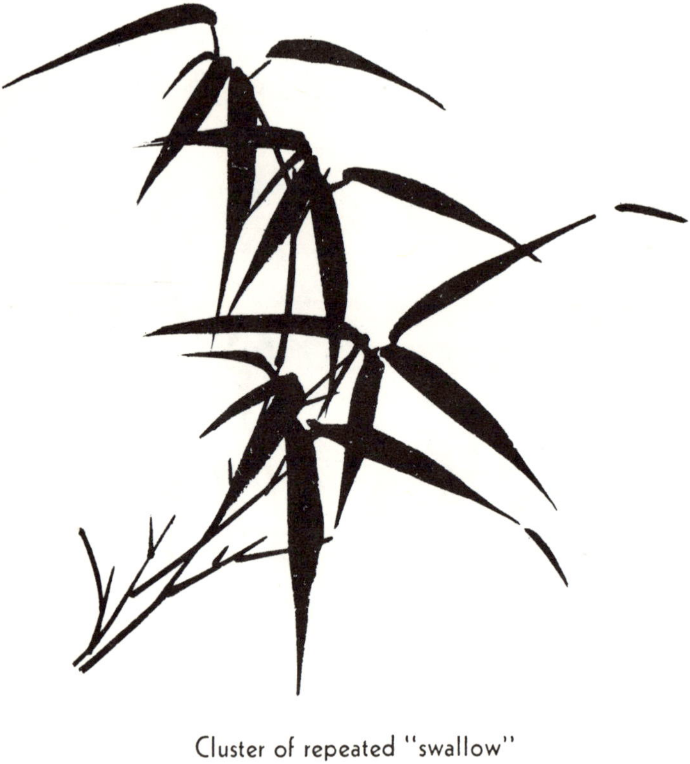

Cluster of repeated "swallow"

According to the masters' books and to what can be seen in all the existing ancient bamboo paintings, there are only five ways of making clusters of leaves, namely the five presented above. And the special purposes for these clusters, as taught by the masters are as follows:

(a) The cluster of repeated "fish tail" is for the tip of hanging branches growing over a wall or a rock.

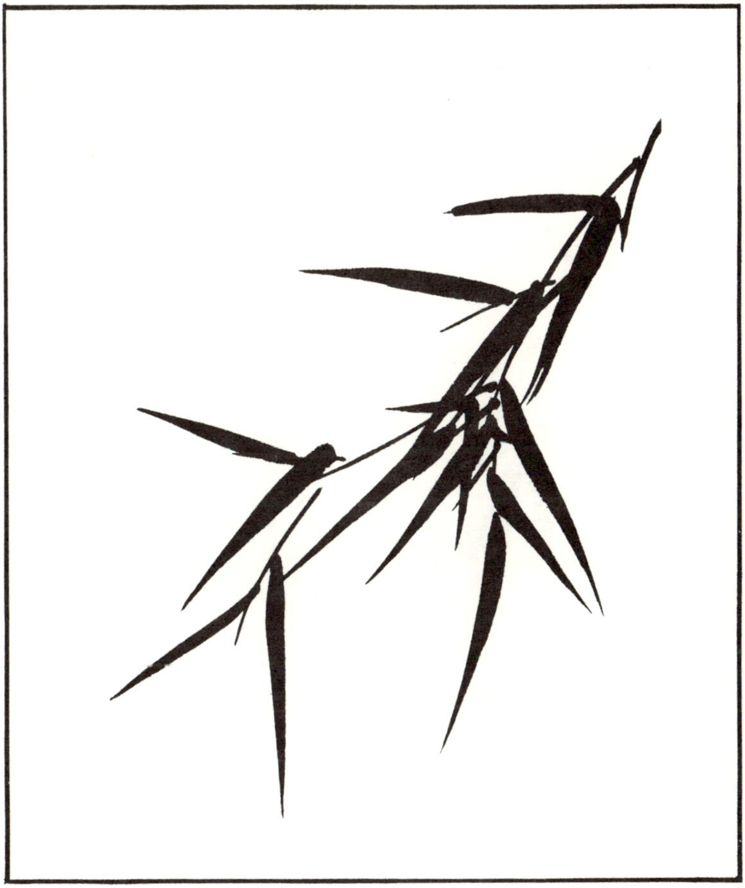

(b) The cluster of repeated "double fishtail" is for a branch of young bamboo growing upward or horizontally.

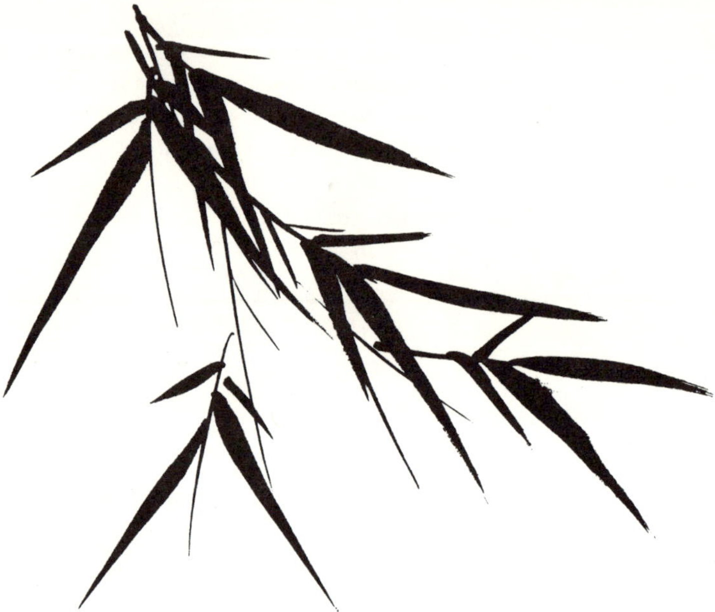

(c) The cluster of repeated "between" is for old leaves hanging loosely at the top of the bamboo plant.

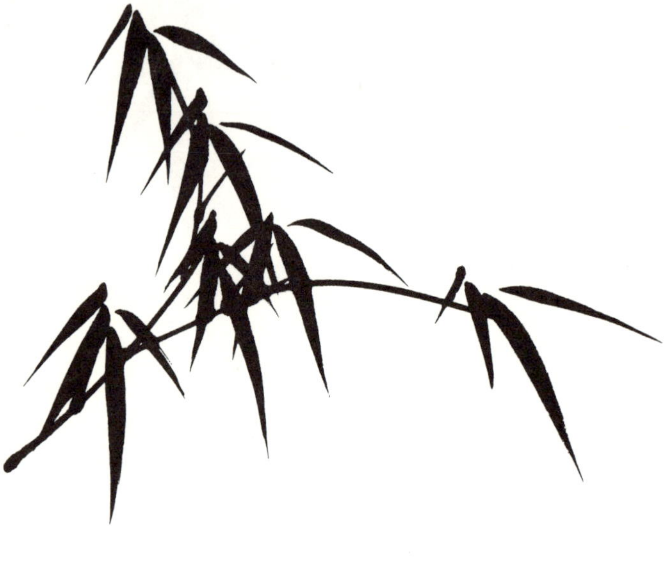

(d) The cluster of repeated "division" is for heavy leaves at the top of a bending bamboo plant.

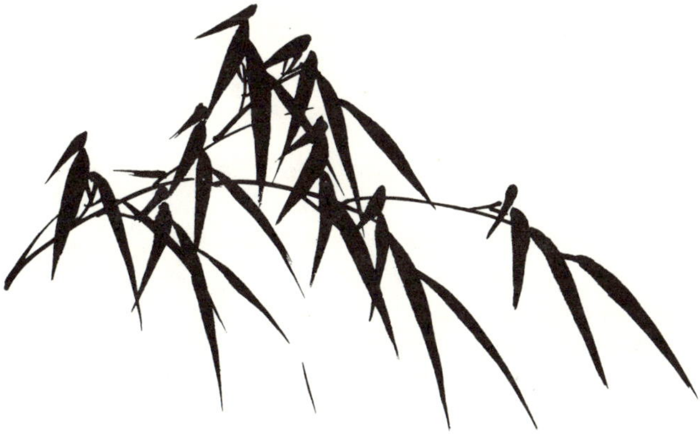

(e) The cluster of repeated "swallow" is for the bamboo in the rain or loaded by dew drops early in the morning.

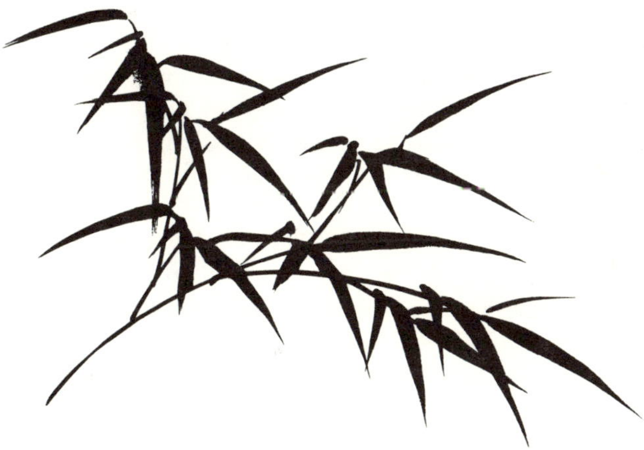

C. A study on the overlapping of leaf strokes

We know already that for making leaves, we must have the brush fully saturated with thick ink. It is because the strokes for leaves are broader and the movements are faster. When the ink is not enough, our leaves will have split endings and the strokes will be dry and broken.

When we put one "leaf-formation" after another on each tip of the "horn," due to the varied distances among the tips, the overlapping of leaf-strokes is inevitable.

When a certain part of our rice paper—like the place where one heavy leaf-stroke is just made—is already thoroughly saturated and penetrated with fresh ink, it is not wise to add immediately any other stroke across or even too close to it. Since there is not much room for more ink, the overlapping stroke will force the excess ink to spread outward and make both strokes swollen and gross. So, remember, no overlapping of wet and fresh strokes!

We have discussed the leaves done in thick black ink. How about different shades of leaves? Gray colored leaves not only create the feeling of distance, but also break the monotony of values. When we paint bamboo in color, we can put different amounts of black or gray into colors to produce gradations of shades and tones. Colors also can be thinned into tints just as the black can be diluted into grays.

First, we saturate the brush, which has been rinsed well in clean water, with medium gray or the color desired. Then, put black or darker color on the tip of the brush, and make strokes one after another without changing the condition of the tip of our brush, that is without dipping into the stone or adding color and water. The shades or tones of all the strokes made by us will show the gradation of value in such sequence:

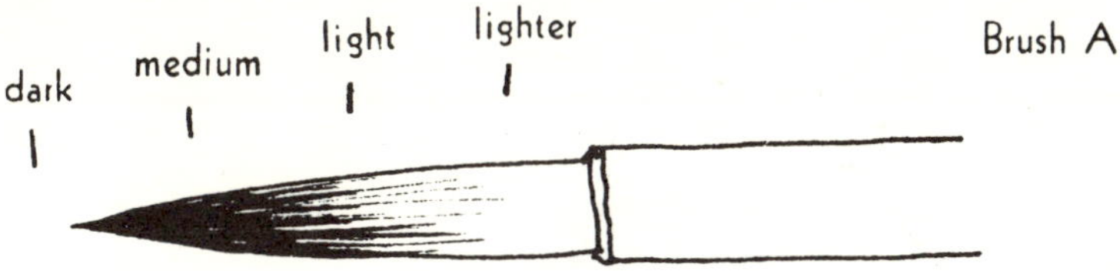

When we put more black or darker color on the tip, the scale of the shades or tones will be:

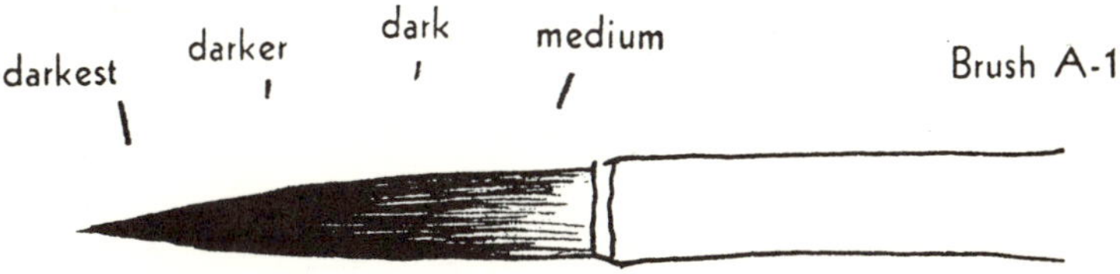

When less black or lighter color is added on the tip, the scale of values will be:

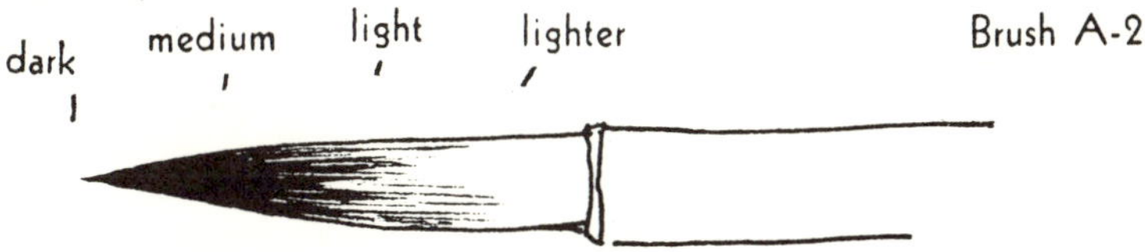

If we have black or darker color already on the tip, we may put drops of clean water on the tip or at the side to produce lighter or thinner tints.

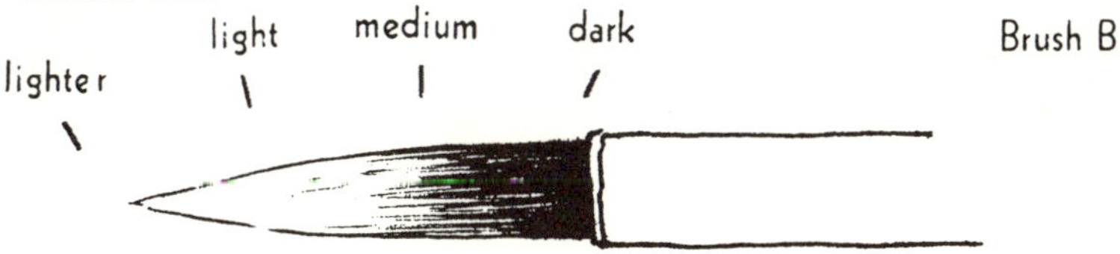

And in ths same manner, we also may put black or dark color at the side of a light color or a clean brush, to produce the sequence as below:

Brush C

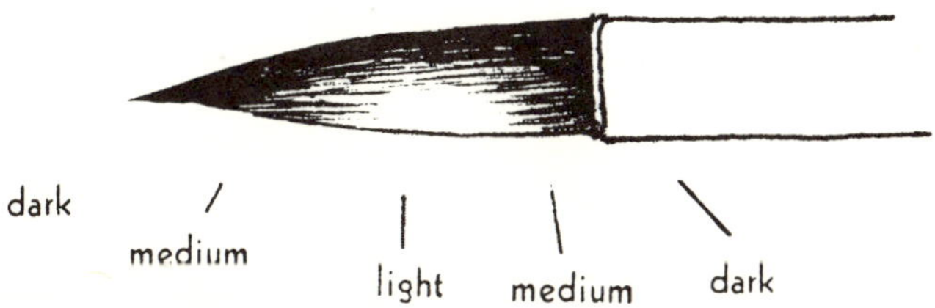

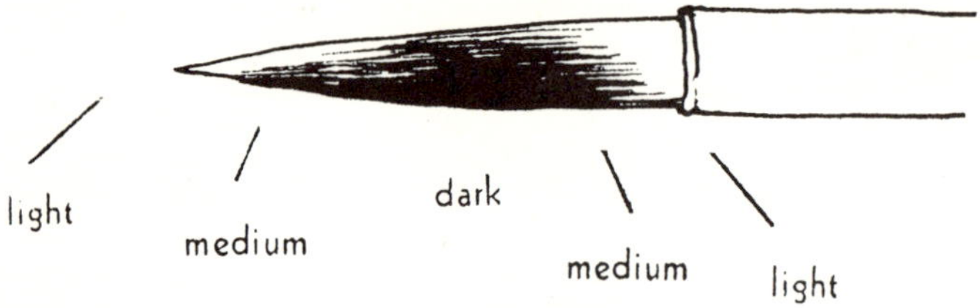

Brush D

Since we must avoid the immediate overlapping, we paint with black and darkest color first, separately but accordingly. After the black or the darkest color on the tip is used up, we have the lighter and thinner shades or tones. Then we have strokes in gray and lighter colors for the leaves behind and in the distance. Black and darker colors are thicker and dry easily; while the lighter ink and color are more transparent. So all the light color strokes can be placed freely afterwards, overlapping the black or darker strokes.

However, lighter color always means thinner color in this type of painting. It runs and spread much faster than black and darker colors. So, when a lighter stroke meets other lighter strokes, precaution must be taken to prevent any smearing or running of the ink or the tearing off of the heavily darkened paper by our moving brush.

For the leaves at the b ack of the trunk with the inside or underside facing the viewer, we prefer to have them in gray or lighter color. By using a piece of thicker paper to protect the trunk from staining or unwanted overlapping, we can make complete leaf-strokes freely and continuously. This procedure is similar to that of putting a dark stem behind a light one.

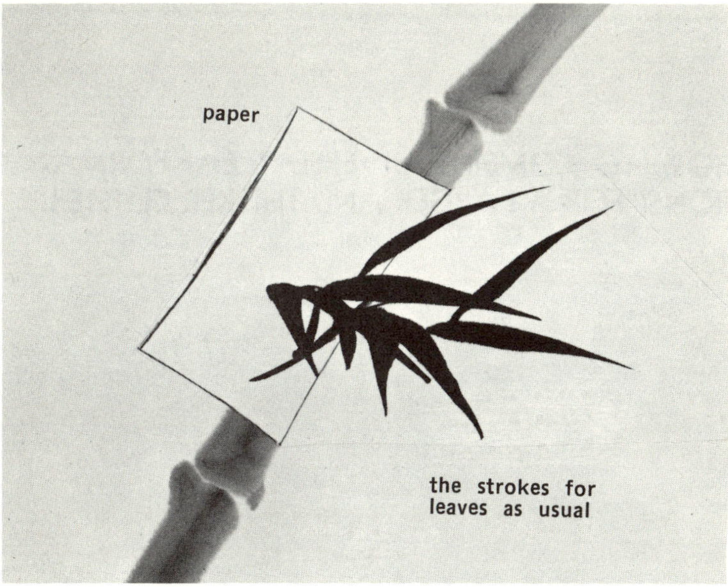

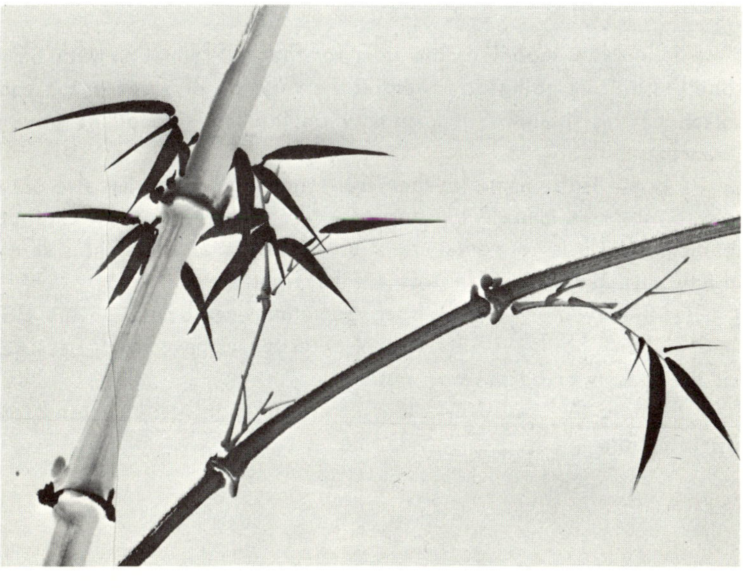

after the paper is removed

HOW TO COMBINE DIVERSE "LEAF-FORMATIONS" FOR A LARGER AND THICKER CLUSTER.

A. **The combination of "new moon," "boat" and "fish tail"**

1. "New moon" is the term for the leaf-formation with only one brushstroke going downward at the very top of a tree or the end of an exterior branch. It is usually on the left side of the young branch.

"Boat" is the name for the "leaf-formation" which has also only one brushstroke made at the top of a tree or the end of an exterior branch. But it is horizontal and it usually starts at the right side of the young branch and moves to the right.

The "new moon" and "boat" sometimes can be put on the old and broken "horns" to suggest that other leaves have fallen because of the season or because of a storm.

In these two cases, no other "leaf-formation" will be used for combination.

2. "Fish tail" and "goldfish tail" formations are for the younger leaves growing upwards. The Book of Bamboo of the **Mustard Seeds Garden** called these two together with "new moon" and "boat" as the "leaves elevating."

Here is an example to show how we may arrange all the "leaves elevating" in one cluster.

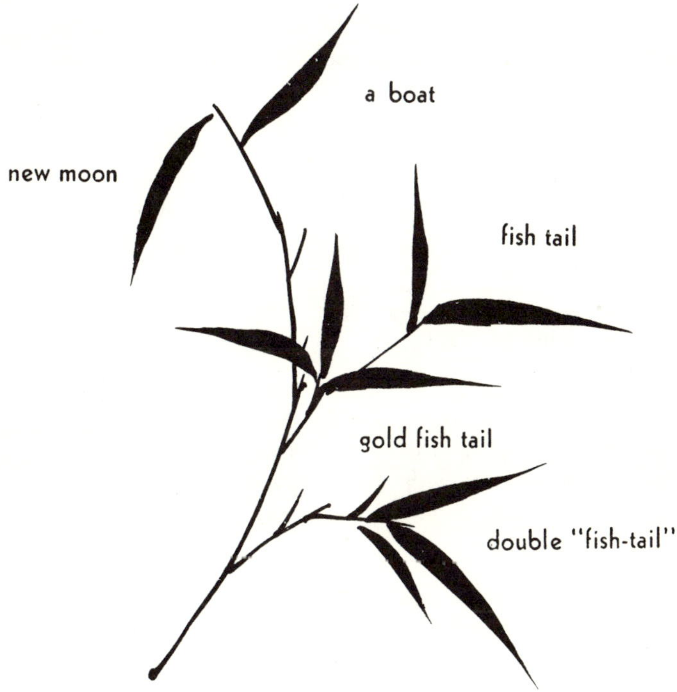

This combinations always suggests the branch of a young bamboo or the top of a bamboo standing erect and gayly in the morning sun.

B. "Double fish tail" and "between" combination

The leaf-formation of "double fish tail" looks like an inverted "between," however a "between" is not an upside-down "double fish tail." The following comparison shows their difference :

Double fish tail	Between
(1) The "double fish tail" belongs to the group of "leaves elevating."	(1) The "between" belongs to the group of "leaves pendant."
(2) Each pair of leaves is similar to the "fish tail." The outer pair of leaves splits with a wider angle than that of the "between."	(2) The upper pair of leaves hanging down on both sides makes a rather narrow angle, and the other two in the center are even closer.
(3) The outer pair of leaves are always made by broader and longer strokes and the other two are smaller.	(3) Strokes for the two pairs of leaves are almost equal in size.
(4) We can place them going downward, if the branch is long enough to bend itself downward.	(4) Less upward "between" should be made, even on the upward "horns," this formation always drops vertically from the top of the "horns" and looks downcast and long-faced.

Here are examples:

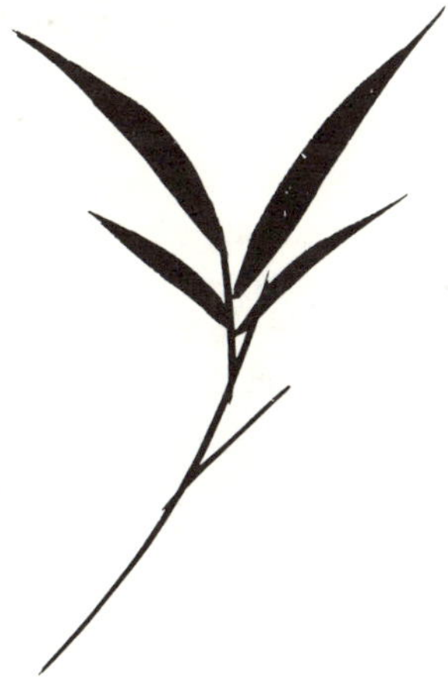

"between"

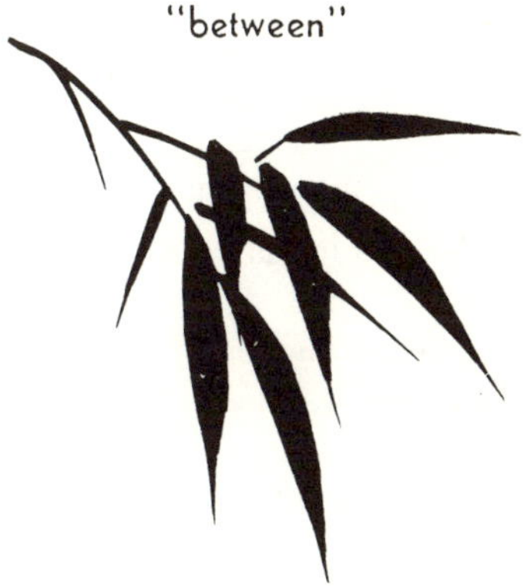

C. "Swallow" and "division" combination

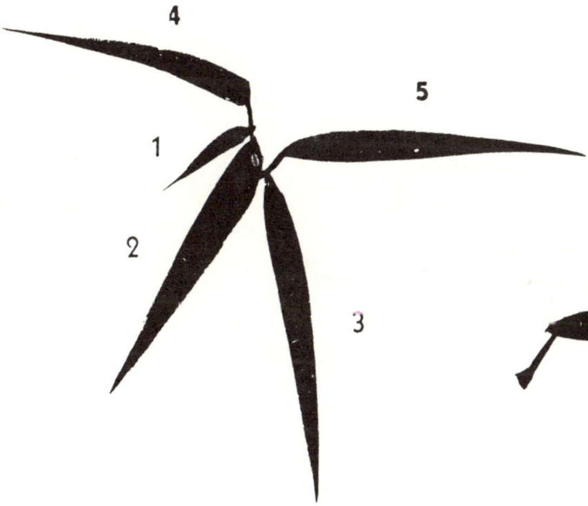

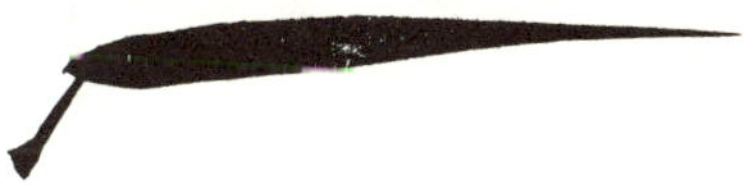

The "swallow" looks like the top view of a raised "between," it is always found on top of a tree or an erect branch. It resembles a swallow flying away from the viewer: #1 stroke is for the head of the swallow, #2 and #3 are the tails and #4 and #5 are the stretching wings. Many artists prefer to repeat the "swallow" formation to suggest thick clusters of leaves which are extending forward and toward the viewer. Branches underneath which are mostly foreshortened and covered by leaves become less important and even unnecessary.

Since the trunks or big branches are the determining factors or a whole composition, we must finish them first in making our bamboo painting. Therefore all the "swallows" in front of these trunks and big branches which have been finished first, should be done in the darkest color or in total black and directly on them to cover them up.

"Landing bird" and "frightened bird" are developed from the "swallow." These two can be applied alternatively at the same places where the "swallows" are desired.

To avoid too much overlapping, some strokes of these "birds" can be eliminated. We called them "broken" leaf-formations.

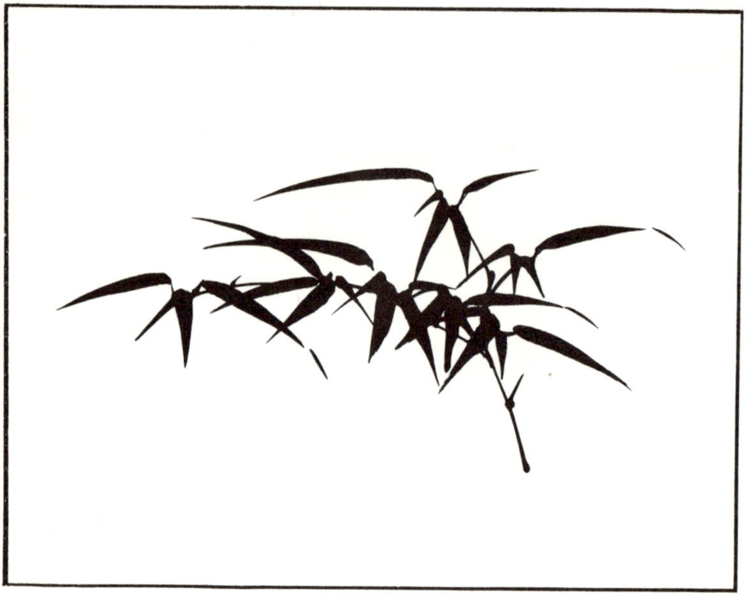

We have studied the relationship between the leaves hanging downwards in front of the "horns" and the side-view leaves spreading away from the trunk. And so we know front-view leaves can be either the "between" or the "double fish tail," but the side view ones are always the "division."

The construction of "division" is similar to that of the Chinese character **fen** in **ts'ao** style, which means division, hence the name.

Since the "swallow" is the top view of a raised "between" and the "division" is somewhat like the side view of a "between," so the relations among them can be seen simply in the illustration below:

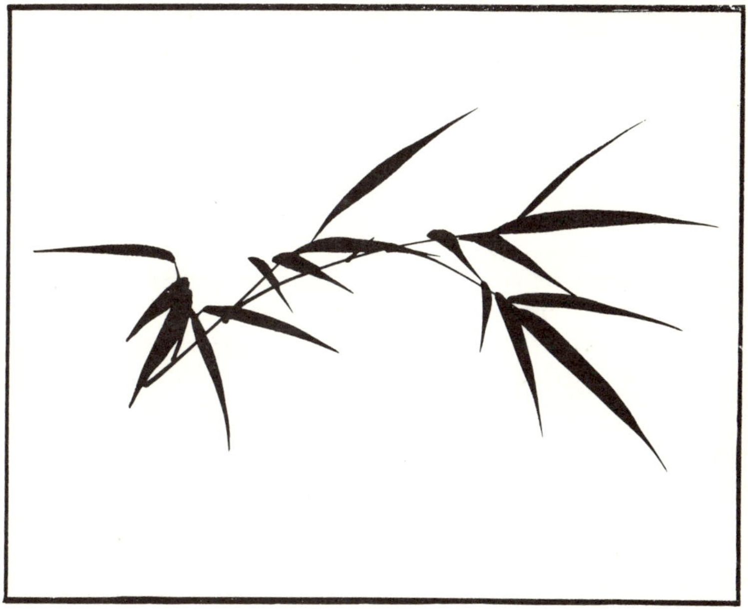

Based on the principle of linear perspective, our "swallows" can "fly" towards any directions to show the mass and the depth of each cluster of leaves. There is no definite way of deciding whether a "between" has turned enough to the side and should be called a "division" already. When a "swallow" which is actually a raised "between," or a pendant "between" gradually turns more and more to the side, and lies horizontally to the viewer, it becomes a "division" too. The only difference is that this kind of "division" is lifted by a rising branch.

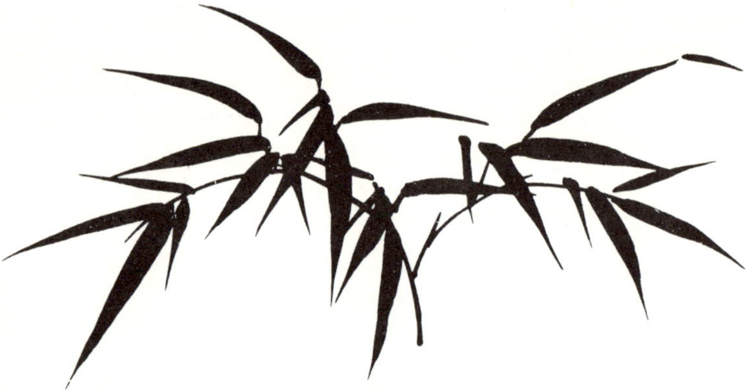

When there are many "swallows" gathered on top of an erect branch or "horn," they always "fly" towards the center of the cluster of leaves. They are arranged in such a way as to form a shape resembling an umbrella.

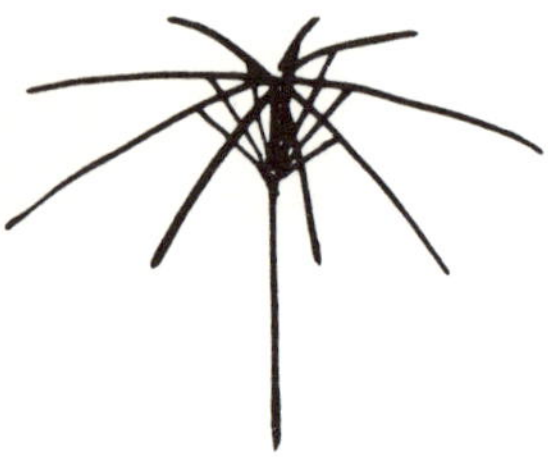

abstraction

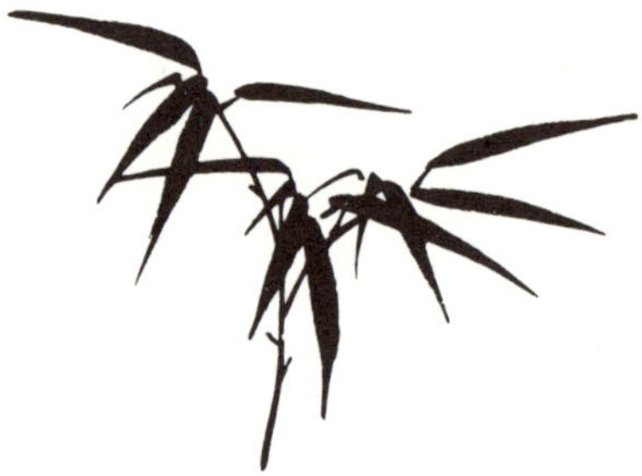

a resembling of an umbrella

Following are more examples. Notice the alteration of different "birds." Practice the "broken birds" casually and spontaneously, taking care not to break the rules of peispective.

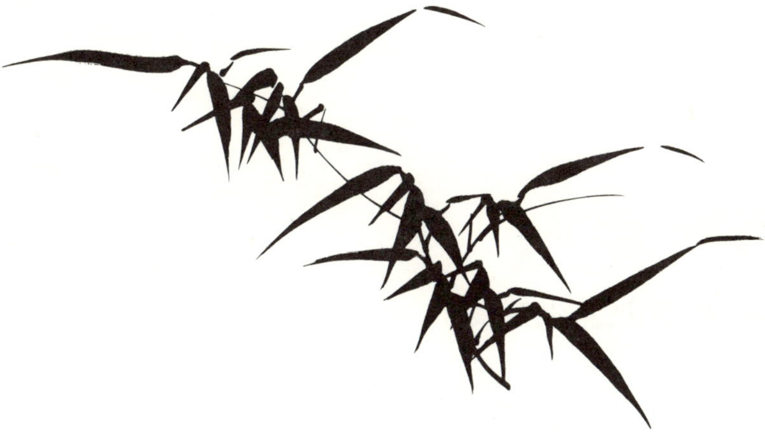

"broken birds"

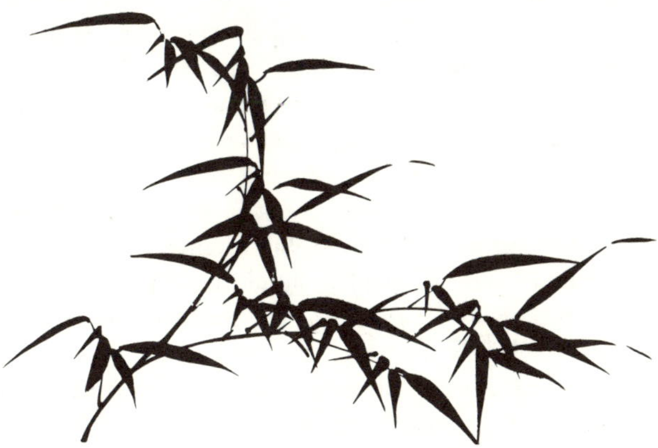

another group of "broken birds"

THE TRUE ESSENCE

Countless varied compositions may be produced by combining different trunks—old, broken or young, with roots exposed, or with shoots, in different seasons—with different leaves—leaves in the rain, leaves in the wind or breeze, with dew or snow, in the morning or afternoon, under the moonlight. Furthermore there are numerous techniques to produce the above effects. To describe all of these would take another book, which I am planning to write. Many of these techniques were discovered centuries ago by masters to achieve their expressive purposes. But no one can say that there are no more novel ways to be found by us in the future. However, no matter what method or tool is used, the essence of Chinese brush-work, i.e. "painting" and calligraphy, should not be forgotten. The Chinese artist never really "paints" in the Western sense; rather, he "writes" different subjects the way he writes in the different styles of calligraphy. The strokes of calligraphy are used to "paint." Taking advantage of especially designed tools for photographic realism is only for the artisan, and considered for many centuries as poor taste too.

In this concluding part of this introductory book of a planned series of manuals on Chinese painting, I would like to emphasize the characteristics of Chinese painting. A Chinese artist seeks to show by subtle use of his brush the most salient characteristics of his subject. He attempts to suggest the true essence of his subject with the greatest simplicity.

Chinese painting is primarily studio work. No artist paints directly from his subject. He often sketches until he feels his brush has mastered the intricacies and only then turns to compose the painting, which is executed quickly and surely from memory; or having made mental notes after close observation of a chosen subject, the artist retires to his studio and puts together a painting which to him, expresses a typical aspect of the subject. Thus no Chinese landscape is an accurate delineation of any one geographical location, but rather an assemblage of many aspects which have been observed by the artist in his travels in a particular vicinity. No portrait is solely a realistic likeness of the subject; instead it is also a visual expression of character. No bamboo painting is a photo-tographic copy of some bamboo at some particular place, seen from a particular angle; instead it is a suggestion of the true essence of the bamboo, an expression of the qualities of a true Chinese gentleman, whom the bamboo symbolizes.

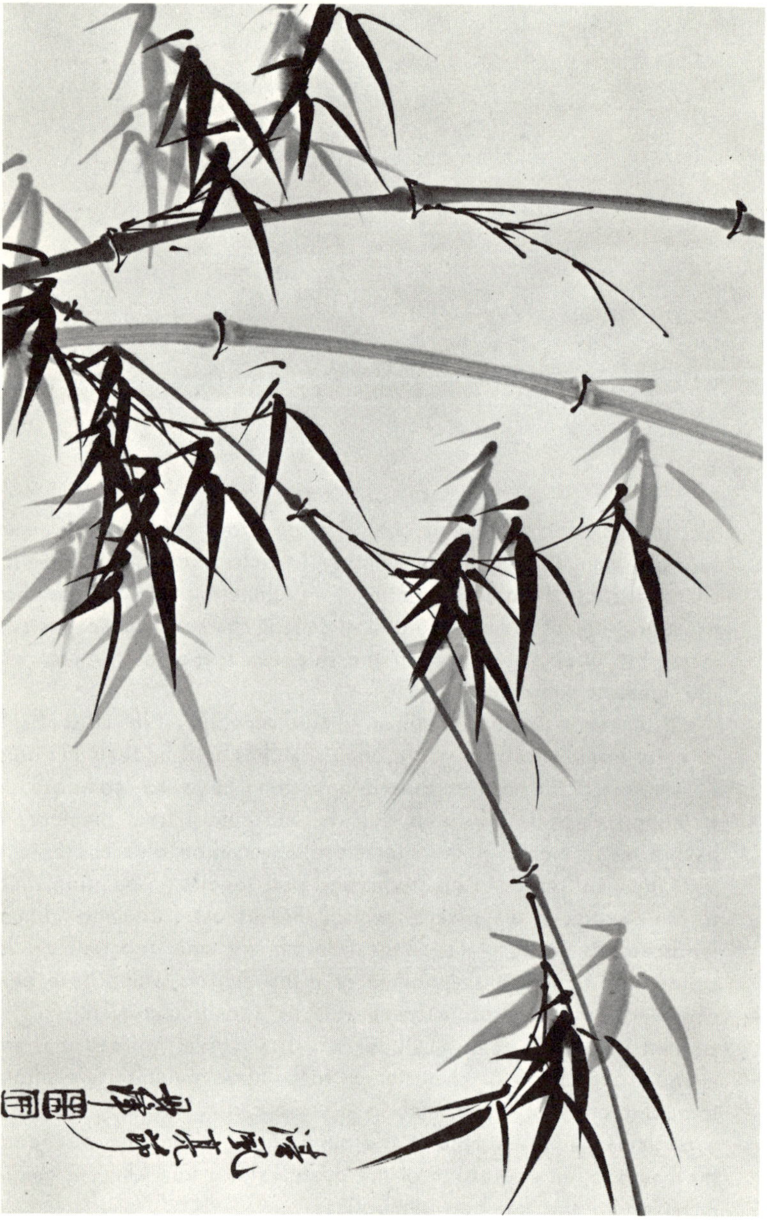

Bamboo in the wind by I-Hsiung Ju

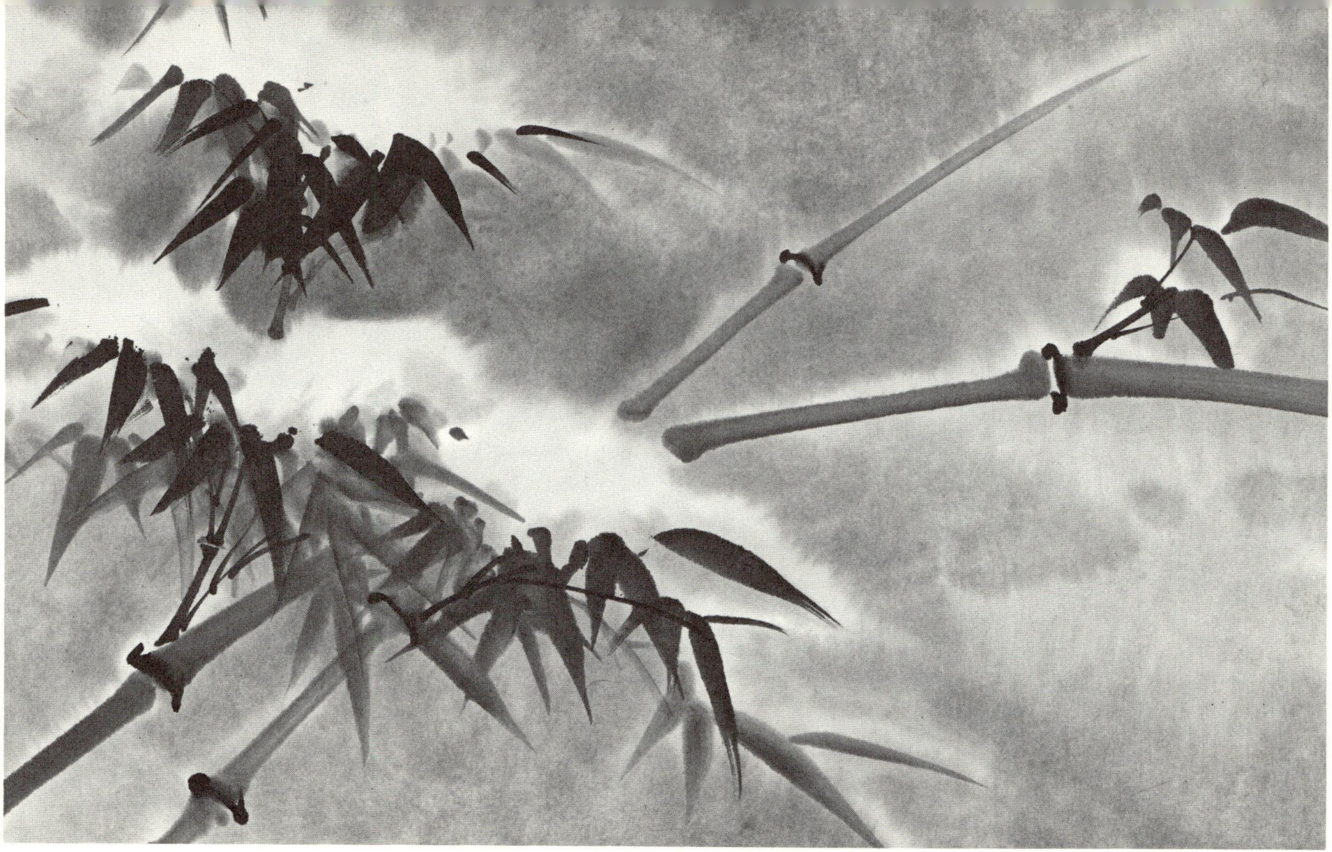

Bamboo under the snow

by I-Hsiung Ju

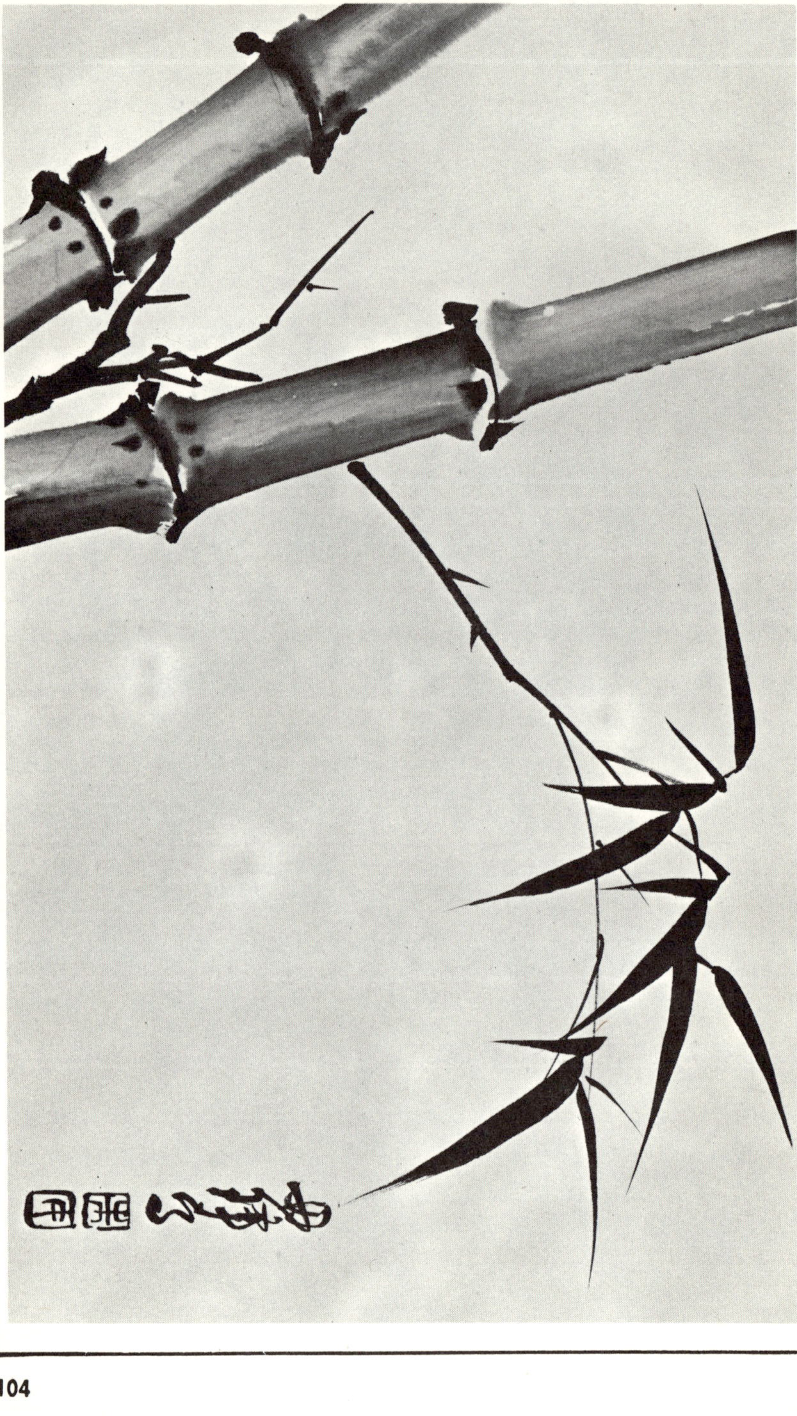

Old bamboo by I-Hsiung Ju

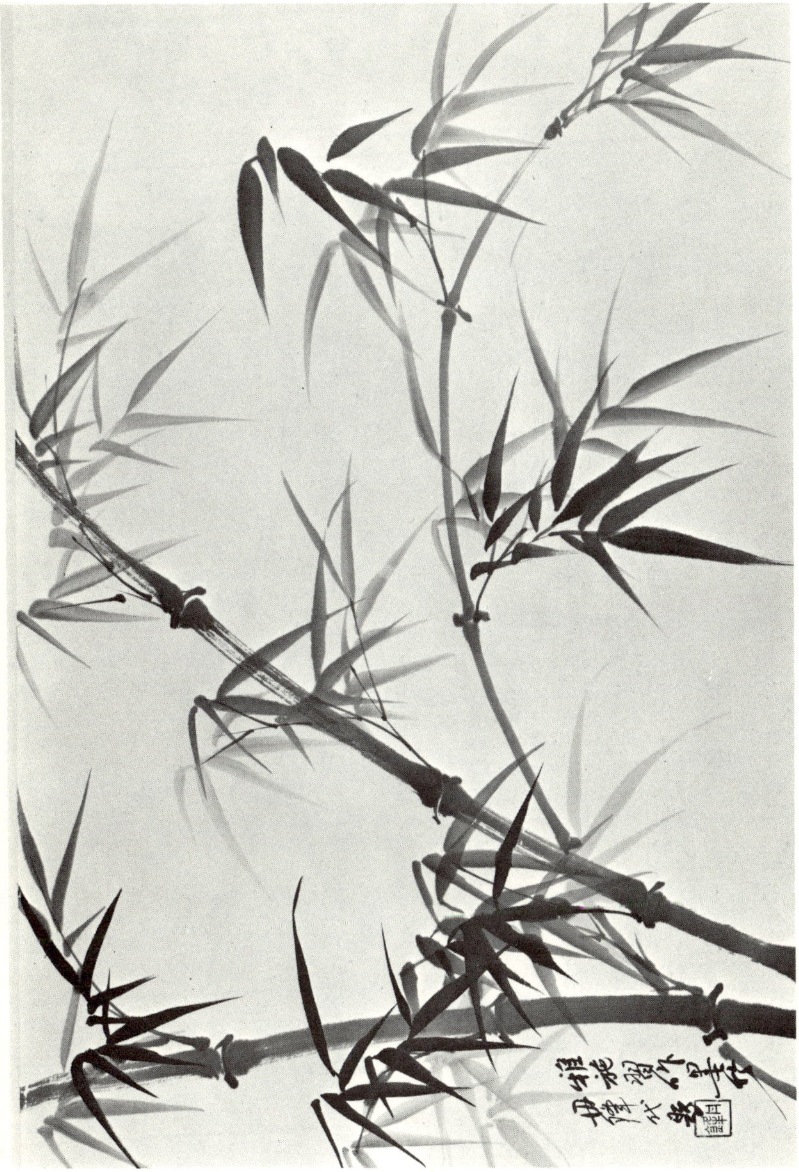

AFTER RAIN

By Grace C. Ju

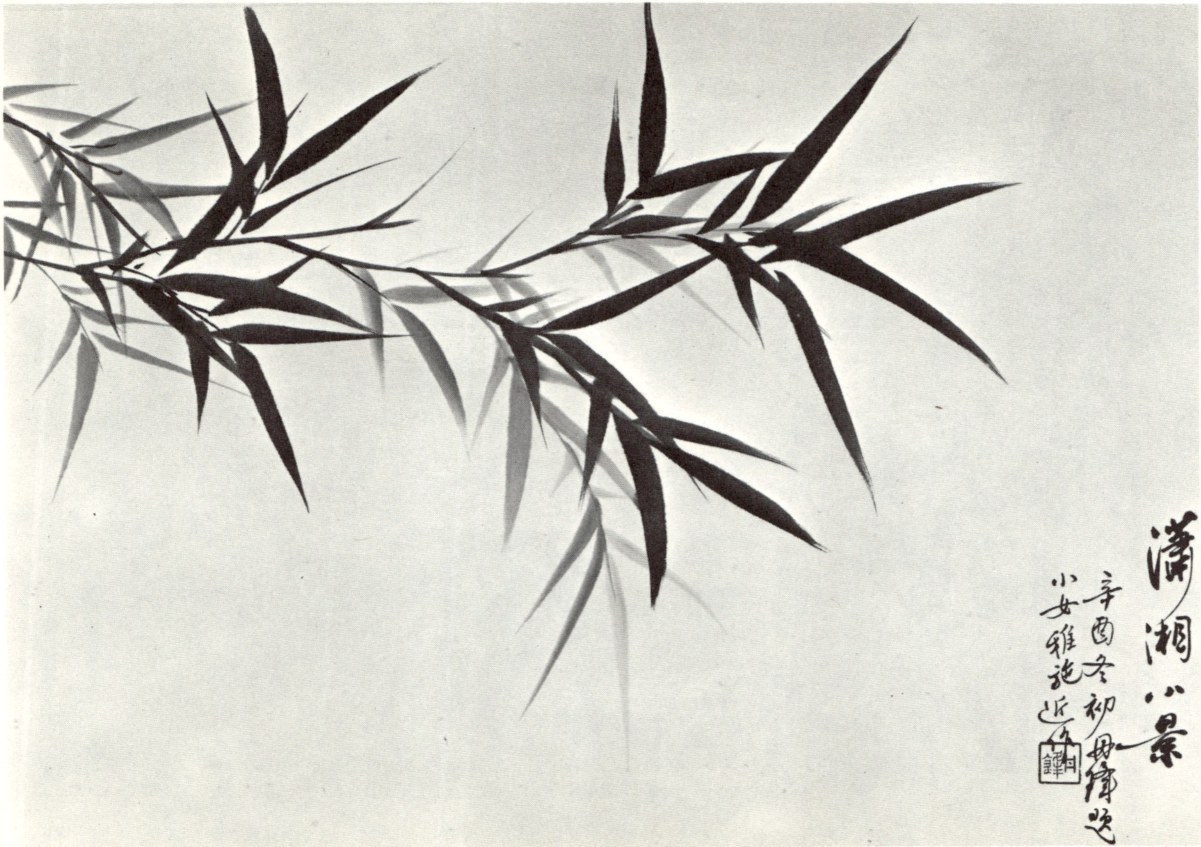

SPRING

By Grace C. Ju

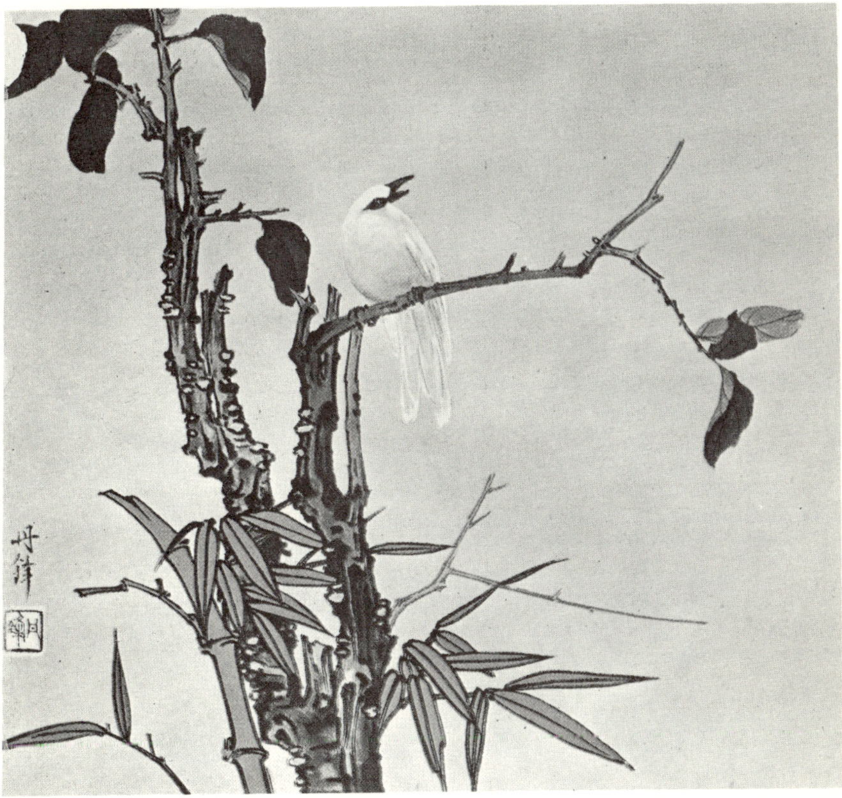

RED LEAVES (Silk) By I-Hsiung Ju

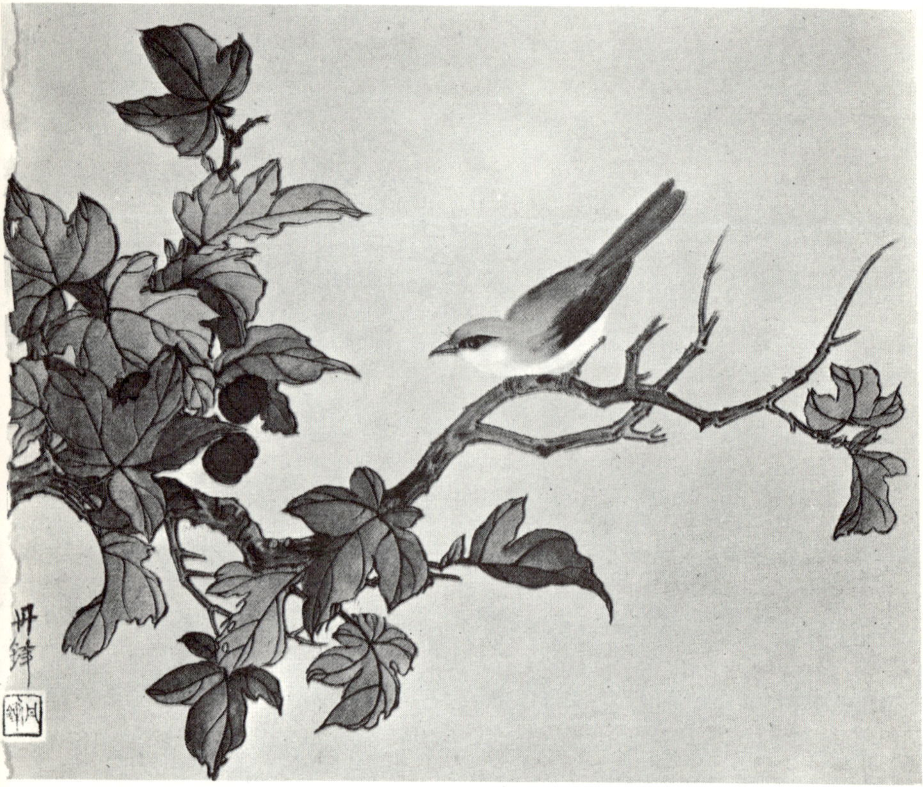
YELLOW BIRD (Silk) By I-Hsiung Ju

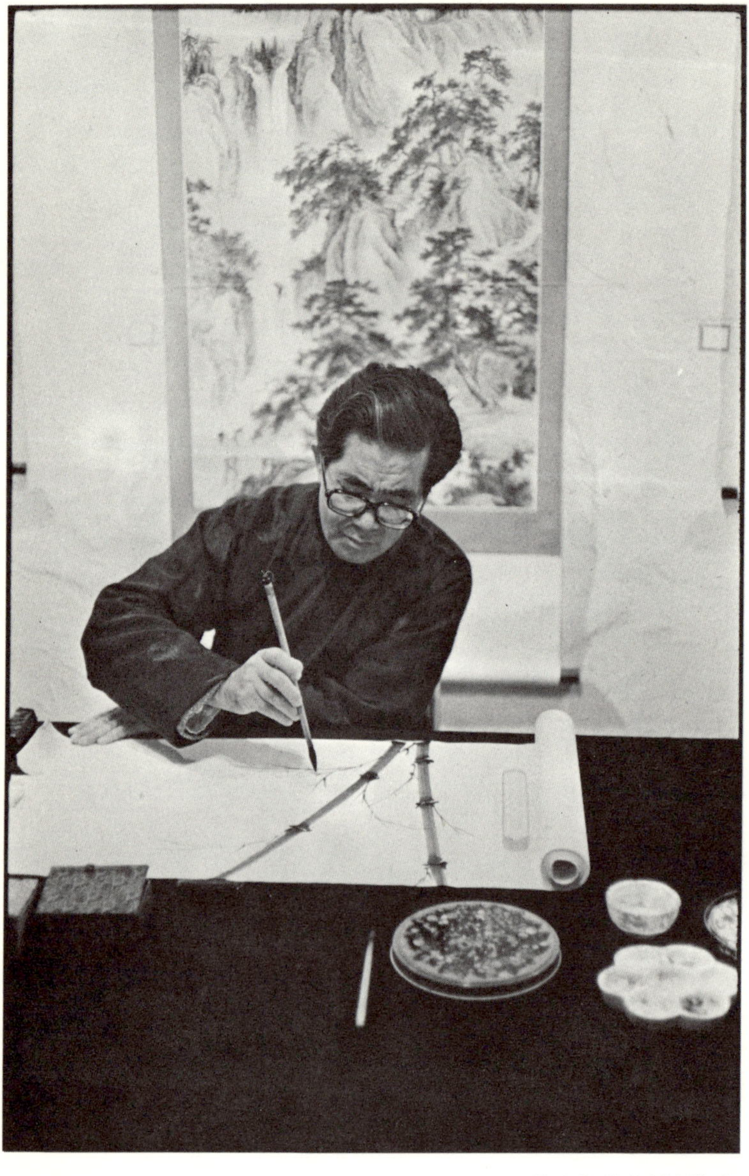

Photo By W. Patrick Hinely

I-Hsiung Ju

Prof. I-Hsiung Ju was born in Kiangyin, Kiangsu, China. 1923. He graduated from the National University of Amoy in 1947 and received his A.B. degree in Chinese Art and Literature. Because of the war in China, he went to the Philippines to teach and at the same time to continue his studies. He graduated from the University of Santo Tomas in Manila and received his B.F.A. and M.A. in History.

He has been considered as one of few Chinese artists able to blend two worlds of style, technique, and idiom to produce a unique form of painting which is both modern and traditionally Oriental. According to him, "a Chinese artist is not only a painter, but also a poet and a philosopher."

A prize winner in graphic art, oil painting and Nanga works in various countries, a writer and a lecturer on art, he has held many one-man shows in Australia, Canada, China, England, Hong Kong, Japan, the Philippines and the United States. He is the author of many books and numerous papers on Chinese art.

As an Artist-in-Residence and Professor of Art at Washington and Lee University, he was selected Professor of the Year for 1971 by the Ring-tum Phi, the Washington and Lee University newspaper and was awarded the Best Art Educator of the Year for 1974 by the Chinese National Writers and Artists' Association in Taipei, Taiwan, the Republic of China.